ON-CHIP ESD PROTECTION FOR INTEGRATED CIRCUITS

An IC Design Perspective

THE KLUWER INTERNATIONAL SERIES IN ENGINEERING AND COMPUTER SCIENCE

ON-CHIP ESD PROTECTION FOR INTEGRATED CIRCUITS

An IC Design Perspective

by

Albert Z. H. Wang
Illinois Institute of Technology
http://www.ece.iit.edu/~awang

KLUWER ACADEMIC PUBLISHERS
Boston / Dordrecht / London

Distributors for North, Central and South America:
Kluwer Academic Publishers
101 Philip Drive
Assinippi Park
Norwell, Massachusetts 02061 USA
Telephone (781) 871-6600
Fax (781) 681-9045
E-Mail < kluwer@wkap.com>

Distributors for all other countries:
Kluwer Academic Publishers Group
Distribution Centre
Post Office Box 322
3300 AH Dordrecht, THE NETHERLANDS
Telephone 31 78 6392 392
Fax 31 78 6546 474
E-Mail < services@wkap.nl>

Electronic Services < http://www.wkap.nl>

Library of Congress Cataloging-in-Publication Data

A C.I.P. Catalogue record for this book is available
from the Library of Congress.

Printed on acid-free paper.

Printed in the United States of America

TO MY PARENTS AND FAMILY

Contents

Acknowledgements

The author wishes to thank all the fellow researchers and IC designers for their shares of contributions to understanding the grand, complex on-chip ESD protection circuit design problem. The book is a selective collection of most of the research advances in the field.

The author acknowledges the sponsorship of a long list of industrial, academic and governmental organizations upon the author's related research activities at the Integrated Electronics Laboratory, Department of Electrical and Computer Engineering, the Illinois Institute of Technology, in various forms.

The author is particularly grateful to his graduate students at the Integrated Electronics Laboratory at IIT for their dedicated and fruitful work on related research topics. Among them are Mr. H. G. Feng for his MS thesis work on mixed-mode ESD simulation-design methodology, Mr. K. Gong for his MS thesis work on ESD-to-circuit interactions and Miss R. Y. Zhan for her PhD dissertation research on ESDcat software development, to name a few. Finally, the author wants to thank Mr. Alexander Greene, Senior Publisher at Kluwer Academic Publishers, for his efforts in producing this book.

Preface

The author first got in touch with the ESD things in late 1996 when he was a Staff R&D Engineer at the National Semiconductor Corporation, where he was assigned by the boss to develop some on-chip ESD protection circuits for mixed-signal IC chips. Ever since then, the not-so-attractive ESD protection circuit design topics have been on and off the author's table both at the company's cubicle and in the office at the Illinois Institute of Technology, where the author joined its faculty team of Electrical and Computer Engineering in Fall 1998. Exactly like all the other IC designers, the author does not really like the little ESD thing and has no fun in doing ESD protection circuit design. Unfortunately, life is life. One has to deal with some unpleasant issues sometime somewhere somehow. The harsh reality is that an IC designer must find the right ESD protection solution for the IC chips. If there is no ESD protection provided, nobody will buy your chips. If there is no sufficient ESD protection for your chips, you will lose the market to your competitors. Period! What makes an IC designer's life even more miserable is the fact that as IC technologies advance, the customer demands for IC ESD robustness and the complexity of on-chip ESD protection circuit design increase dramatically, as evidenced by the huge amount of related papers published in the past decade. While significant progresses have been made in the field of ESD protection research and design, IC designers are deeply bothered by the situation that there are too much qualitative sayings and too few quantitative analyses on the ESD protection design matters. Hence, very little success in predicting ESD protection circuit design is expected. One question commonly heard in the IC circuit design community is that "*how do I design the ESD protection for my chips, with prediction, as a circuit designer?*" This book tries to provide the information necessary to address this tough design challenge from an IC circuit designer's angle.

From IC designers' viewpoint, it is not enough to talk about ESD protection design in the process and device physics language only; and it is unacceptable to continuously rely on the traditional, non-predictive, trial-and-error approaches in ESD protection circuit design practices. Since ESD protection phenomena involve complex multiple level process-device-circuit-electro-thermal coupling effects, the task of ESD protection design is not really about designing a stand-alone ESD protection structure itself. One ought to take into account of the complex interactions between the ESD protection structures and the core circuits being protected in order to successfully achieve full chip level ESD protection. An IC designer ought to be enabled to perform full-scale simulation in order to conduct predictive ESD protection circuit design. With these considerations, the book starts with introduction of ESD protection fundamentals including the ESD origins, ESD test models and standards, etc. It moves on to describe the basic ESD protection devices. ESD protection circuit solutions and advanced ESD protection concepts are then discussed in great details. With these preparations, predictive ESD protection simulation-design methodologies and case studies on the complex ESD-to-circuit interactions are presented. A group of practical ESD protection circuit design examples are provided to enhance the theoretical discussions. Finally, the book concludes with a brief summary on current and future work on ESD protection circuit design. The author wants to point it out clearly to IC designers that, please never try to copy a successful ESD protection circuit, either from this book or anyone else, onto your chips. Most of the time, it will not work for you. It is important to understand that ESD protection circuit design is a custom design and is not portable. It is extremely important to treat ESD protection design as a circuit design task and to conduct ESD design simulation as much as possible. After all, describing a problem quantitatively is more scientific than speaking qualitatively.

An IC designer's best wishes to fellow IC designers: Good Luck!

Albert Z. H. Wang
Illinois Institute of Technology
Chicago, October 2001.

Chapter 1

INTRODUCTION

1.1. A LITTLE HISTORICAL STORY

Electrostatics, or, static electricity, is as old as the time itself. Let us take a quick look into the ancient past. In 600 BC, Greeks discovered static electricity in basic experiments, where they rubbed Amber with a piece of fur and observed attraction of lightweight objects to the Amber. That might not be the very first observation of static electricity generation, however, was believed to be the first documented experiment. Interestingly, the English word "electron" derived from the Greek word "electron" for Amber. Serious work on static electricity might have started with the publication of DeMagnete in 1600 by Gilbert, who also made up the word "electricity". Understandings of modern concepts of static electricity were established after a few more centuries of research activities conducted by Gauss, Coulomb, Faraday, etc. Among them, the famous experiments include that performed by Franklin in 1752, as shown in Figure 1.1, when he flew a kite during a storm and observed the Leyden jar placed close to one end of the kite's wire being charged up; and that conducted by Hertz in 1886, as illustrated in Figure 1.2, where he realized wireless communication by using electrostatic discharge (ESD) devices as transceivers. Interested readers can readily find the detailed information by visiting a Science Museum.

Protection against ESD-caused damages also has a long and rich history as that of taking use of electrostatic discharges. As early as in the fifteenth century, European military agencies started to use a variety of means to handle munitions safely. The lightening rod, invented by Franklin soon after his famous kite experiment, which still benefit us all today, was certainly one of the most significant ESD protection devices in the scientific history.

Figure 1.1 Franklin's kite test in 1752 showing lightening is an ESD event.

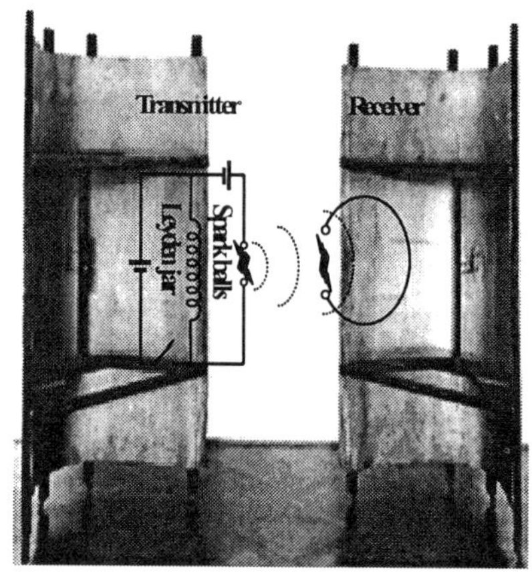

Figure 1.2 Hertz's transmission work in 1886 using ESD spark balls.

Back to today, one could hardly not experience some kind of ESD events in our everyday life. For example, most of us may have experienced the shocking sparks when touching a door handle of a car once and again. One can never image how our everyday life could be without static electricity; meanwhile, one should never overlook the kind of damages caused by ESD as well.

1.2. ESD FAILURE – AN IC RELIABILITY PROBLEM

The ESD problem became a real issue after World War II as highly insulating polymeric materials found widespread usage where substantial static charge accumulation might cause machinery shut downs. However, the devastating ESD damage problem was not taken into serious consideration until the modern microelectronics technologies took the role in our everyday life, which is how ESD failure problem became relevant to semiconductor IC technologies. With the invention of semiconductor transistor in 1947 and development of metal-oxide semiconductor (MOS) technologies in the 1960s, the impact of the invisible ESD phenomena, with a level lower than 1000 volts, became materialized. Some electronic devices can be damaged by an ESD transient of as low as 10 volts. The electronic component and system failures due to ESD events in the electronics industry escalated almost exponentially in the 1970s. Consequently, the military began to develop related standards to govern ESD immunity of electronic products. Particularly, ESD failure is a profound reliability problem to integrated circuits (IC) and poses a grand challenge to the semiconductor industry. This IC ESD failure problem and the ESD protection solution are the topics of this book. As the semiconductor IC technologies advances into the very-deep-sub-micron (VDSM) regime, IC parts become increasingly susceptible to ESD damages. Statistics indicated that up to 30% of all IC failure might be attributed to ESD, as illustrated in Figure 1.3 [1], which costs the semiconductor industry billions of U.S. dollars annually [2]. Such ESD-induced failures are either catastrophic or latent in nature, with the former causes immediate IC malfunction and the latter leads to future failure and lifetime problem. The nature of ESD failures is associated with either high power, i.e., high current, which causes thermal damages to semiconductors and metal interconnect, or, high electric field that ruptures dielectric thin films in ICs. Accordingly, appropriate ESD protection means are developed to protect IC parts against ESD damages.

Figure 1.3 Industrial failure analysis reports substantial IC failure rate due to ESD (data from National Semiconductor Corporation) [1] (*Reproduced here by kind permission of the ESD Association and authors*).

1.3. ON-CHIP ESD PROTECTION - GENERAL REMEDY

The lightening rod invented by Benjamin Franklin in the eighteen century is of course one of the earliest and most efficient protection device against ESD damages. Unfortunately, it does not work for IC parts. In the past three decades, profound progresses were made in understanding the ESD fundamentals relevant to semiconductors. Consequently, there are a long list of means proposed in practices to protect electronic elements and systems against ESD strikes at different levels. At the top level, a systematic ESD control program [3, 4] should be established in manufacturing, transportation and application fields, which consists of preventing electrostatic charging and providing a safe way for static discharging in work environments. Efficient measures include setting up an ESD immunity work place, such as, well-grounded static protective work areas and platforms and ESD protective floor, etc; preventing human body induced electrostatic generation by using ESD protective shoes, clothing and wrist straps, etc; and adopting ESD-combating materials, such as antistatic materials used to prevent charging and static dissipating materials used to safely discharge the accumulated static electricity. Another important factor in ESD control is ESD awareness that include having somebody restlessly yelling "*Caution, ESD!*" in workplaces and promoting use of ESD symbols, as shown in Figure 1.4 [5].

(a) (b) (c)

Figure 1.4 A group of ESD awareness symbols: (a) indicating common point ESD ground, (b) indicating ESD susceptibility, and (c) indicating ESD protection in place.

Developing an efficient ESD control program is not to set up another bureaucratic layer in workplace, there are economic scale benefits spoken by numbers. For example, a failure analysis at Lucent Technologies reports

ESD related manufacturing loses of $325 millions annually before implementing an ESD control program, while effective ESD control brought ESD failure rate down to one part per million units [3]. Now, Stop! "These things have nothing to do with me!" says an IC designer. Well, here comes the opportunity for IC designers to showcase their talents – designing on-chip ESD protection structures. This is the sole topic of this book. The rationale is that, no matter how good an ESD control program could be, ESD occurs. The best way is to protect IC parts at chip level by designing and integrating ESD protection sub-circuits on chips [6]. In early days, Zener diodes were widely used for effective ESD protection. As semiconductor IC technologies advance and ESD protection requirements increase, more ESD protection structures were developed, such as, bipolar junction transistor (BJT) protection, NMOS transistor protection, SCR (silicon controlled rectifier) protection, and primary-secondary dual protection, etc. Today, custom designed ESD protection sub-circuits are commonly used both to accommodate special needs of mixed-signal ICs at deep-sub-micron node and to meet more stringent ESD protection requirements, for instance, higher protection level (e.g., beyond 10 kV human body test model) and extremely fast ESD test model (e.g., the sub-1ns charged device test model). In principle, on-chip ESD protection mechanisms are two-fold in nature: to dissipate the large ESD current transient safely using a low-impedance discharging channel to prevent thermal damages in silicon and/or metal interconnect and to clamp the huge ESD voltage pulse to a safe level to avoid dielectric rupture. The complete ESD protection solution should be realized at full chip level, where the key point is to create an efficient discharging channel from any pin to every other pin on a chip. A main design trade-off is to balance ESD performance and circuit functionality, which are often in conflict due to parasitic effects. Currently, typical ESD protection capacities of 16V/μm for non-silicided grounded-gate NMOS structures [7], 37V/μm for NPN BJTs [8], and 80V/μm for SCR structures [9], all based on human body test model, were reported.

The goal of this book is to provide a thorough reference for practical IC circuit designers to deal with ESD protection design at chip level. With the understanding that ESD protection solutions are normally technology and application specific, and ESD protection design seems always lag in the new chip development chain, this book intends to provide detailed ESD protection fundamentals, and particularly, offer CAD-based practical ESD simulation-design methodologies for circuit designers to conduct predictive ESD protection design concurrently with normal IC design tasks.

1.4. CHALLENGES IN ESD PROTECTION DESIGN

While it is well received that the principle for on-chip ESD protection is to automatically create a shunting path for safe discharge of static electricity, practical design of efficient ESD protection circuits becomes increasingly more challenging as IC technologies continue to shrink. Such challenges in ESD protection design are multi-fold in nature. The first main issue is associated with the complex interactions between ESD protection structures and the circuits being protected, referred to as ESD-circuit interaction in this book. On one hand, performance of ESD protection units are affected by many surrounding factors. A working stand-alone ESD protection unit does not warrant chip level ESD operation because any parasitic devices, both surrounding the ESD protection region and inside the core circuit, has a potential of becoming an unexpected shunting path. Since such parasitic discharge routes are normally not designed to handle an ESD-level transient, pre-mature ESD damages often occur as a consequence even though an individual ESD protection unit might work perfectly by itself. Such phenomena are referred to as circuit-to-ESD influence in this book. On the other hand, inevitably, an ESD protection structure produces parasitic effects, such as, extra parasitic capacitance and ESD-induced noises, which have direct impact on the circuit performance, referred to as ESD-to-circuit influence in this book. Such ESD-induced influences are substantial in many cases, especially in high-speed, high-density mixed-signal (MS) and radio frequency (RF) IC applications in VDSM regime. This problem will only get worse as IC technologies advance. The second major challenge in ESD protection design lie in the rational ESD protection design methodology, essential to ESD protection design prediction. Believe or not, while CAD-oriented design became a common practice in main stream IC chip design, the trial-and-error approach still dominates in today's ESD protection design practices, largely due to the lack of accurate high-current ESD device models and accountable CAD tools for ESD protection design. It is believed that it usually takes average two to three times for an experienced designer to complete a successful ESD protection design. This certainly has negative impacts on cost reduction and time-to-market of new chip development. The third key challenge is related to the non-portability of ESD protection solutions, not only because ESD protection structures are geometry and technology dependent, but also because ESD solutions are product-specific. A successful ESD protection unit for one chip usually does not work for a different type of chips designed in the same process, not to mention in different new process technologies. Together, it makes ESD protection design not so much a fun work. Consequently, ESD protection design is often blamed for failure of a new chip, which is actually due to the separation of IC circuit design and ESD protection design, most likely.

1.5. SCOPE OF THIS BOOK

This book is written by an IC circuit designer, who has suffered a lot in new chip development due to the ESD problems, for practical IC designers. Therefore, it is developed from a circuit designer's perspective and focuses on circuit design issues. Many real world design examples are included to ensure the usefulness of this book.

After the introduction in Chapter 1, fundamentals of ESD phenomena and existing ESD testing models are discussed in Chapter 2, including various ESD testing standards established and used in the field, as well as commonly used commercial ESD testers. Chapter 3 discusses basic device physics of typical semiconductor devices used as ESD protection building elements ranging from diodes and bipolar junction transistors (BJT), to metal-oxide-semiconductor field effect transistors (MOSFET) and silicon-controlled rectifiers (SCR). It also explains the on-chip ESD protection principles. Chapter 4 covers more advanced ESD protection circuit solutions suitable for VDSM IC applications including input protection, output protection, power clamps, and whole-chip ESD protection schemes. Chapter 5 discusses the new frontier in ESD protection design for mixed-signal and RF ICs, including developing novel, low parasitic, compact ESD protection structures for VDSM applications. ESD failure mechanisms and failure analysis (FA) techniques, which are essential to ESD protection design successes, are discussed in Chapter 6. In Chapter 7, practical ESD protection circuit design layout issues are covered that often play decisive roles in final ESD protection circuit functionality at chip level. This chapter also discusses the relationship between ESD protection circuit design and process technologies, including process tweaks for optimal ESD protection, the trade-offs, and ESD protection for emerging technologies. ESD device modelling and ESD simulation techniques are discussed in Chapter 8. Practical ESD simulation-design methodologies, aiming to ESD protection design prediction, will be described in depth in this chapter. In Addition, a number of practical ESD protection circuit design cases are presented in this chapter. In Chapter 9, the complex ESD-circuit interactions will be discussed using real world design examples. Conclusion remarks and future consideration on ESD protection circuit design are given in Chapter 10. Finally, a set of ESD protection design checklist, designed to assist IC designers in handling with real ESD protection projects are given in the Appendix C.

REFERENCES

[1] Merril, R. and Issaq, E., "ESD Design Methodology", *Proc. EOS/ESD Symp.*, 1993, pp. 233-237.

[2] Green, T. J. and Denson, W. K., "Review of EOS/ESD Field Failures in Military Equipment", *Proc. EOS/ESD Symp.*, 1988, pp. 7-14.

[3] Dangelmayer, G. T., *ESD Program Management*, Boston, Kluwer Academic Publishers, 1999.

[4] *ANSI ESD S20.20-1999: Standard for the Development of an ESD Control Program*, The ESD Association, Rome, NY, 1999.

[5] *ANSI ESD S8.1-1993: ESD Awareness Symbols*, The ESD Association, Rome, NY, 1993.

[6] Amerasekera, A. and Duvvury, C., *ESD in Silicon Integrated Circuits*, New York, Wiley, 1995.

[7] Wang, A., Tsay, C., Lele, A. and Deane, P., "A Study of NMOS Behaviour under ESD Stress: Simulation and Characterization", *Microelectronics Reliability, 38,* Elsevier Science, 1998, pp.1183-1186.

[8] Chen, J., Zhang, X., Amerasekera, A. and Vrotsos, T., "Design and Layout of a High ESD Performance NPN Structure for Submicron BiCMOS/Bipolar Circuits", *Proc. EOS/ESD Symp.*, 1996, pp. 227-232.

[9] Wang, A. and Tsay, C., "On a Dual-Direction on-Chip Electrostatic Discharge Protection Structure", *IEEE Trans. Elec. Devices*, *Vol. 48, No. 5,* April 2001, pp. 978-984.

Chapter 2

ESD TEST MODELS

2.1. NATURE OF ESD PHEMONEMA

Let us start with basic concept – static electricity. Repeating the ancient Greek game, i.e. to rub a piece of amber against a piece of fur, positive and negative electric charges are generated and stay in the amber and fur separately. These generated electric charges, normally remaining in the generating objects, are called electrostatic charges. In fact, electrostatic charges can be created whenever any two objects touch and then separate from each other, with the degree of static electricity generation depending upon the materials involved. A charged object will form an influential space around it called electric field and will affect any other charged entity in its neighbourhood via its electric field. The measure of such electrostatic influences, called Coulomb's Force, is governed by the Coulomb's Law,

$$F = \kappa \frac{Q_1 Q_2}{r} \tag{2.1}$$

where F is the Coulomb's Force, Q_1 and Q_1 are the amount of electrostatic charges in the each objects involved, r is the distance in between, and κ is a constant. Consequently, it requires certain amount of energy, called work, to move a charge in an electric field, which is characterized by electrostatic potential at related points in the electric field.

The most common electrostatic charge generating procedure is called triboelectric charging, where two objects are brought into contact and then separates from each other. Electrostatic charges of opposite polarities are generated and remain in each entity accordingly to its nature, leading to different electrostatic potentials for the two objects. The tendency of such electrostatic generation depends upon the inherent properties of the materials involved. As indicated by a typical Triboelectric Series in Table 2.1 [1, 2], some materials tend to retain positive charges, while others prefer to generate negative charges. The further apart the two involved materials in the triboelectric series, the stronger the electrostatic charging could be. Other electrostatic charging mechanisms include field induction, ion beam charging, and photoelectric charging, etc.

Table 2.1 A typical Triboelectric Series

Materials	Electrostatic Polarity
Air	+
Acetate	↓
Glass	
Human hair	
Nylon	
Wool	
Fur	
Lead	
Silk	
Aluminium	
Paper	
Polyurethane	
Cotton	
Wood	
Steel	
Hard rubber	
Acatate fiber	
Nickel, copper, silver	
Brass, stainless steel	
Synthetic rubber	
Polyurethane foam	
Saran wrap	
Polyester	
Polyethylene	
Polyvinylchloride (Vinyl)	
Silicon	
TEFLON	
Silicone rubber	-

When two objects with different electrostatic potentials are brought into close proximity, transfer of electrostatic charges between the two objects occurs. This process is called **electrostatic discharge** (ESD). While ESD is a rather general concept and ESD phenomena occur almost everywhere, the ESD event relevant to semiconductor ICs is generally referred to as electrostatic discharge with a short duration of about 150 ns, which generates very high current and voltage transients, up to a few tens of Amperes and kilovolts. Table 2.2 shows a good ballpark picture for the severity of electrostatic charging in an electronic manufacturing environment [1]. Such fast and large ESD transients may cause severe damages to IC parts.

The electronic components that are susceptible to ESD damages are referred to as ESD sensitive devices, often called ESDS elements. ESD damage to an ESDS component is classified by its ability to survive ESD transients, a property referred to as ESD sensitivity. To increase IC parts' ESD survivability, dedicated on-chip ESD protection structures are used to protect IC chips against ESD stresses. The ESD protection performance level, often regarded as ESD robustness, is determined by the ESD failure threshold voltage of an IC part, normally called ESDV in units of volts or kilovolts (kV). In order to estimate the ESDV level of an IC part, or to specify its ESD rating on data sheet, one needs to test the IC part by stressing the device under test (DUT) with emulated ESD zaps, a procedure called ESD zapping. To obtain reliable ESD zapping results, it is extremely critical to be able to generate reproducible ESD transients using an ESD tester satisfying accepted ESD specifications. Numerous such ESD test models have been developed as standards, both military and commercial, which are discussed in the following sections.

Table 2.2 Example data of electrostatic charge generation in workplace [1] (*Reproduced by kind permission of Kluwer Academic Publishers and T. Dangelmayer of Ion Systems*)

Motions	Electrostatic potential at relative humidity	
	10%	55%
Walking across a carpet	35 KV	7.5 KV
Walking across a vinyl floor	12 KV	3 KV
Worker on bench	6 KV	0.4 KV
Removing DIP[1] parts from plastic tubes	2 KV	0.4 KV
Removing DIPs from vinyl trays	11.5 KV	2 KV
Removing DIPs from polystyrene foam	14.5 KV	3.5 KV

[1] DIP = dual-in-line package.

2.2. HBM MODEL

The most well understood ESD mechanism is associated with human body caused electrostatic discharge, which is characterized by a human body model (HBM). The HBM ESD event describes a discharge procedure where a charged human body contacts a device directly and electrostatic charges transfer from the human body into the device. The electrostatic transient generated therefore stresses the component. It is fairly easy to understand that absolutely repeatable HBM ESD events are impossible, because every human body has different electrical characteristics, so does each component under test. Nevertheless, average HBM ESD situation can be generalized as a typical HBM ESD event, based upon which several standardized test models were developed. One of the earliest and most widely accepted HBM test model is the military standard, MIL-STD-883E Method 3015.7, published in 1989 [3]. This military test model uses a simplified equivalent circuit, illustrated in Figure 2.1, to describe the HBM ESD event, where a charged human body is modelled by a charged human body capacitor, C_{ESD}=100pF, and typical human body discharging resistance is specified as R_{ESD}=1500Ω. Table 2.3 lists typical parametric elements of the HBM model

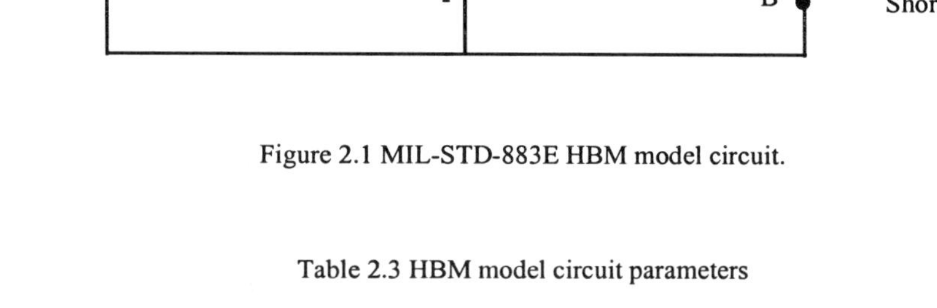

Figure 2.1 MIL-STD-883E HBM model circuit.

Table 2.3 HBM model circuit parameters

Elements	Specifications
R_0	$10^6 - 10^7$ Ω
R_{ESD}	1500Ω
C_{ESD}	100pF ±1%
S_1	HV relay

circuit. A HBM ESD event is then simulated as electrostatic discharge from a pre-charged C_{ESD} into an ESDS DUT via R_{ESD}. To ensure reproducible HBM discharge waveforms and reliable HBM ESD testing data, the MIL-STD-883E defines a standard short-circuit HBM ESD output waveform as in Figure 2.2, with typical parameters listed in Table 2.4. A HBM ESD tester can adopt different model circuit, however, it must be able to deliver a discharging waveform fitting with that given in Figure 2.2 (solid line) and Table 2.4, verified by a short-circuit discharge under a V_{ESD}=4000V. HBM zapping tests of a device are performed by stressing recommended pin combinations three times, each using both positive and negative pulses, with a minimum zap interval of 1 second. The ESDV levels of ESDS components are classified into three categories as listed in Table 2.5. Importantly, MIL-STD-883E sets detailed specifications for a HBM ESD zap waveform, to which a HBM model ESD tester must conform, regardless how a tester is constructed. However, in real applications, parasitic parameters in a tester may have substantial influences on the output ESD waveforms produced during ESD zapping tests. Even the parasitic effect of a DUT device could impact the ESD stressing waveform, especially when testing large high-pin-count chips [4-6]. The HBM ESD model circuit shown in Figure 2.1 is certainly over-simplified, because it delivers an instant discharge waveform, showing in dashed line (L_{ESD} = 0H) in Figure 2.2, that does not even conform to the standard HBM ESD waveform shown in solid line in Figure 2.2, which has a finite rise time, t_r, due to parasitic inductance in discharge path. A modified HBM model circuit is shown in Figure 2.3 where L_{ESD} reflects parasitic inductance (typically 6 to 10μH) in discharge channel. An analytical model can be obtained by solving a second-order Kirchhoff current loop equation for this circuit, which is depicted by

$$i_{ESD}(t) = \frac{V_{ESD}}{2L_{ESD}\sqrt{\alpha^2 - \omega_0^2}}(e^{S_1 t} - e^{S_2 t}) \quad (2.2)$$

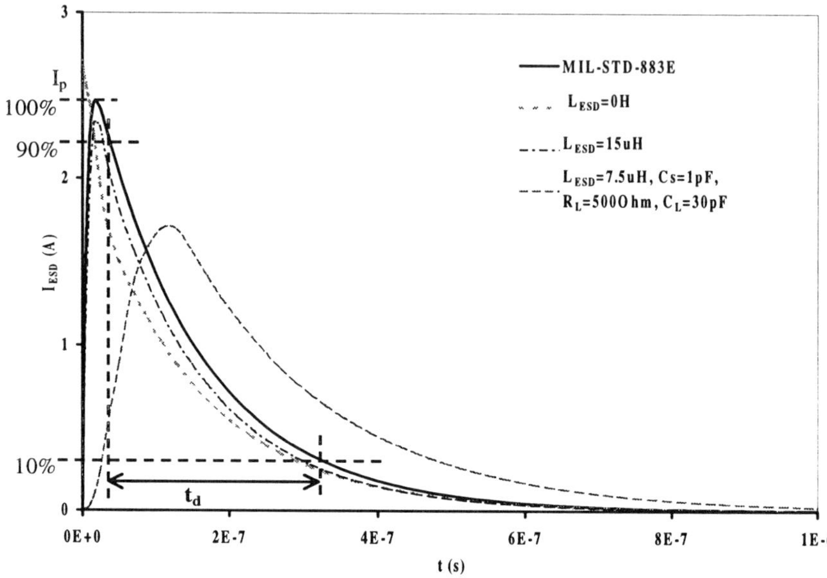

(a)

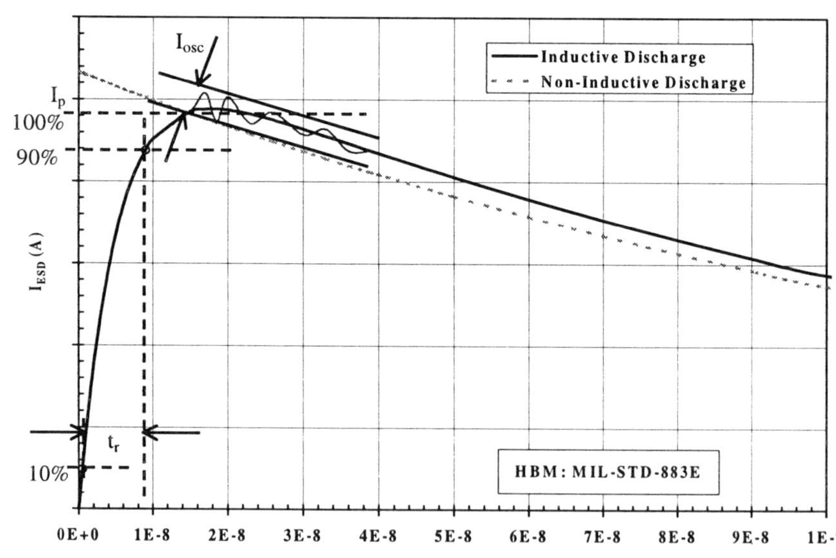

(b)

Figure 2.2 MIL-STD-883E specified HBM ESD waveform. The solid line is the waveform defined by the standard, while the dashed lines reflect parasitic effects in HBM testers.

where i_{ESD} is the discharging current, V_{ESD} is the ESD potential, damping factor $\alpha \equiv (R_{ESD} + R_L)/2L_{ESD}$, natural frequency $\omega_0^2 \equiv 1/\sqrt{L_{ESD}C_{ESD}}$, and $s_{1,2} \equiv -\alpha \pm \sqrt{\alpha^2 - \omega_0^2}$. Since HBM ESD fits with an over-damped discharge procedure, $\alpha > \omega_0$ must hold, requiring $L_{ESD} < 56.26$ μH. Typically, $L_{ESD} = 1.5 \sim 15$μH for $t_r = 2 \sim 20$ns is expected with $L_{ESD} = 7.5$μH corresponding to $t_r = 10$ns. Of course, lots of bells and whistles can be added on to account for other parasitic effects in real ESD measurements. For example, a fourth-order model circuit is illustrated in Figure 2.4, where C_S is stay capacitance in a human body and C_{DUT} accounts for parasitic capacitances of both a test board and an ESDS DUT device [4, 5], for which a simplified analytical fitting equation is available [6]

$$i_{ESD}(t) = V_{ESD}C_{ESD}\{\frac{\omega_0^2}{\sqrt{\alpha^2 - \omega^2}}\}e^{-\alpha t}\sinh\{\sqrt{\alpha^2 - \omega_0^2}t\} \qquad (2.3)$$

These parasitic effects make it extremely challenging to design reliable ESD testers for accountable ESD tests because typical parasitic parameters in a HBM ESD tester may lead to significant deviation in the ESD tester waveform as depicted in Figure 2.2a (dashed lines). To address these uncertainties, improved HBM ESD test standards were proposed by several organizations. For example, ESD STM5.1 from the ESD Association [7],

Table 2.4 Short-circuit HBM ESD waveform specifications

Parameters	Specifications
t_r (Rise)	< 10ns
t_d (Decay)	150 ± 20ns
I_{osc} (Oscillation)	< 15% I_p (peak)

Table 2.5 MIL-STD-883E HBM ESDV classifications

Classes	ESDV levels
Class 1	< 2KV
Class 2	2 KV – 4KV
Class 3	> 4KV

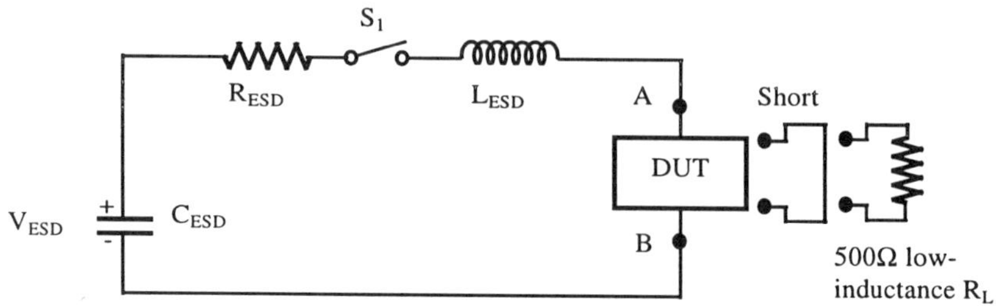

Figure 2.3 A modified 2nd–order HBM ESD model circuit includes parasitic L_{ESD}.

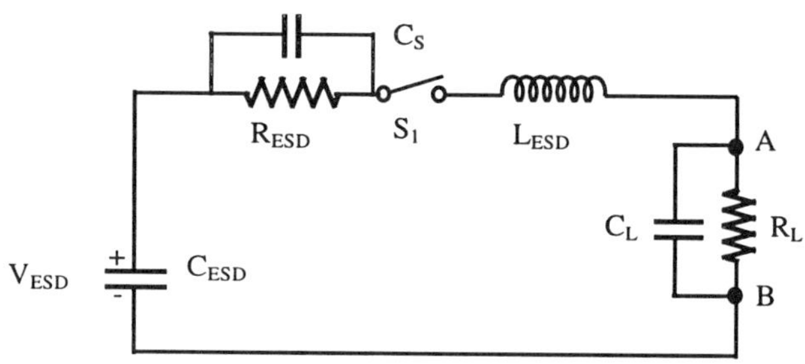

Figure 2.4 A 4th–order HBM model circuit includes more parasitic elements.

JEDEC JESD22-A114-B from the Electronic Industries Alliance [8], and AEC-Q100-002 from the Automotive Electronics Council [9]. Other than targeting for different applications, a main common modification is an extra tester verification step using a 500Ω load in addition to short-circuit calibration. This 500Ω-load calibration is used to ensure that the ESD zap waveform conforms to the standard HBM ESD waveform specifications even under a sizable parasitic C_{DUT}. A detailed comparison table for different HBM standards is included in the Appendix A. It is obvious that none of the standards can guarantee an absolutely reproducible ESD zap waveform and all the models specify acceptable waveform envelopes.

2.3. MM MODEL

In addition to human body caused ESD events, any charged object can discharge into an ESDS device when touching it as discussed in Section 2.1. The most common ESD events of such kind, as far as IC parts are concerned, occur when metallic machinery contacts IC components in an IC manufacturing environment. Typical such actions include in-line inspection and ATE (automatic test equipment) tests. What is unique here is that, unlike in a HBM ESD case, the parasitic resistance is very low for metallic machinery. Consequently, the peak ESD current becomes much higher than that produced in a HBM ESD event. A new ESD test model, called Machine Model (MM) was therefore developed, initially by the EIA in Japan as a worst-case HBM model [10]. Several modified MM ESD test standards have been published, including ESD STM5.2 from the ESD Association [11], JEDEC JESD22-A115-A from the Electronic Industries Alliance [12], and AEC-Q100-003 from the Automotive Electronics Council [13], to name a few. A MM ESD model circuit is similar to that for HBM model. For example, the AEC-Q100-003-Rev-E specifies an MM model circuit as illustrated in Figure 2.5 with C_{ESD} = 200pF, negligible R_{ESD} and L_{ESD} in discharge path, which must be calibrated in both short-circuit and 500Ω load

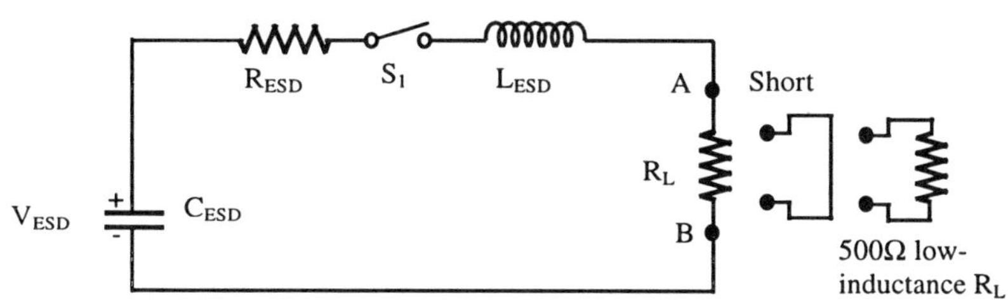

Figure 2.5 A MM ESD model circuit including parasitic elements.

conditions to ensure its output ESD waveforms fitting with the curve defined in Figure 2.6 (solid line) and specified in Table 2.6. Because of its negligible discharge resistance, a MM ESD waveform produces very high peak currents and is oscillatory in nature due to parasitic inductances in testing systems, which certainly cannot be predicted by the over-simplified model circuit given by the standard. However, an under-damped second-order model circuit with non-zero L_{ESD} and total discharge resistance, defined as $R = R_{ESD} + R_L$, as shown in Figure 2.5, can be used to describe the MM ESD waveform. An analytical model can then be obtained as

$$i_{ESD}(t) = \frac{V_{ESD}}{L_{ESD}\omega_d} e^{-\frac{R}{2L_{ESD}}t} \sin \omega_d t \tag{2.4}$$

where $\omega_d \equiv \sqrt{\omega_0^2 - \alpha^2}$ and $\omega_0 > \alpha$. However, although everything specified, it is actually extremely difficult to build up an MM tester that can supply a standard ESD pulse waveform faithfully and reproducibly. The root cause is that its discharge resistance is very low and the parasitic parameters, from both the test board and the ESDS DUT devices, can influence the output waveform dramatically. For example, as shown in Figure 2.6 (dashed lines), a small amount of inductance and resistance in the discharge path, e.g., L_{ESD} = 1uH and R=1Ω, will cause a sizeable deviation from the standard waveform; while a suggested Philips model [4] with L_{ESD} = 2.5uH and R=25Ω gives a totally different ESD waveform. For readers' convenience, a comparison table for different MM standards is given in Appendix A.

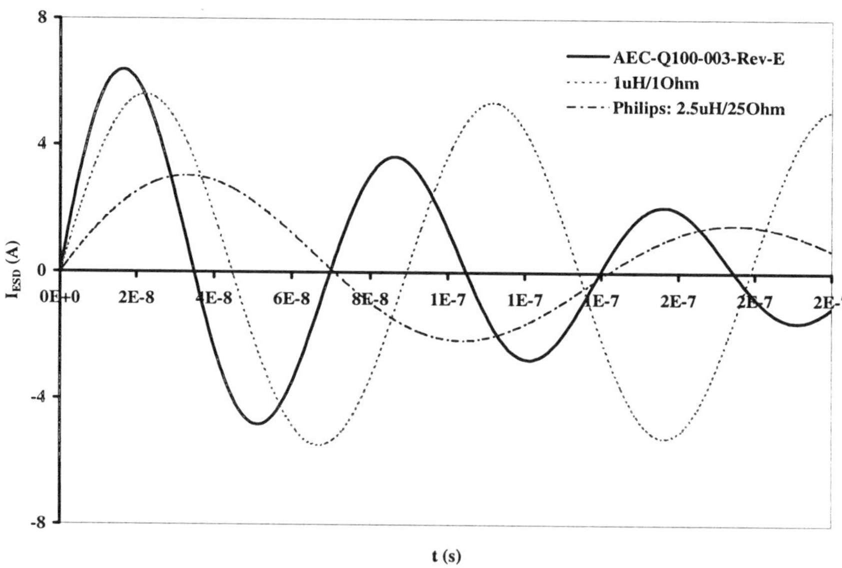

Figure 2.6 An AEC-Q100-003-Rev-E specified MM ESD pulse waveform (solid line). Real world ESD pulses delivered by MM testers (dashed lines) may be quite different due to parasitic effects in discharge channels.

Table 2.6 Short-circuit MM ESD waveform specifications

V_{ESD} (V)	1st (+) peak I_p (A)	2nd (+) peak I_{osc} (A)	Main period t_p (ns)
400	6.0-8.1	67-90% of I_p	66-90

2.4. CDM MODEL

Charged device model (CDM) depicts an ESD event, different from that of HBM type, where a charged IC part discharges when a pin contacts a grounded object or conductive surface. The CDM pre-charging occurs often in manufacturing and field application environments, for example, when sliding a part down from a feeder in an automated assembler or a shipping tube. A CDM ESD event is different from that of HBM in nature because CDM is a self-discharge procedure. However, though charges are stored in a relatively small parasitic capacitance of the ESDS DUT device, a CDM discharge is very rapid and produces very large ESD current (up to several tens of Amperes) because of extremely low discharge resistance and inductance. Therefore, a CDM ESD event could be very deadly and becomes more a concern as IC technologies advance, because the ever-thinner dielectric films in CMOS technologies (< 75Å) are more prone to CDM ESD damages. Currently, as human body related ESD immunity gets improved consistently, CDM ESD event are getting more attentions.

Although the CDM concept was introduced a long time before [14, 15], it is extremely difficult to build up a CDM ESD tester because parasitic elements in the discharge path, depending upon DUT device sizes and discharge heads, have tremendous influences on the CDM ESD pulses produced. Two general types of CDM ESD testing apparatus are available currently. One is based on the generic CDM test model where an ESDS DUT device is tested directly, called a non-socketed CDM tests. The other method places a DUT component into a socket and charges and discharges the DUT via the socket – called socketed CDM tests, which is also termed as socketed CDM (SDM) [16]. There are several existing CDM ESD testing standards available, for example, ESD STM5.3.1 from the ESD Association [17], JEDEC JESD22-C101-A from the Electronic Industries Alliance [18], and AEC-Q100-011 from the Automotive Electronics Council [19], etc. A CDM ESD test model is illustrated in Figure 2.7, where an ESDS DUT part is placed upside down on a charge plate, charged by a high-voltage source, and then discharges through a discharge head. There are two different means for charging: direct charging and field induction charging. The former method, as shown in Figure 2.7a, charges the DUT by direct contact. However, one has to be cautious not to damage the IC part in charging it – a procedure similar to HBM discharge. The latter means, as shown in Figure 2.7b and recommended by many test standards, charges the DUT device via electric field induction to avoid possible charging damages. To ensure pulse reproducibility, the discharge head shall have negligible parasitic parameters, i.e., $R_{ESD} = 1\Omega$ and L_{ESD} is as small as possible. An equivalent circuit for the CDM model is given in Figure 2.7c. Because of its self-discharge nature, the C_{ESD} of a DUT is very small. Taking the JESD22-C101-A as an example, $C_{ESD} = 6.8\text{pF} \pm 5\%$. A little work is needed for one to derive an analytical formula for the CDM ESD pulse waveform based upon the second-order network, which is the same as that for MM ESD model.

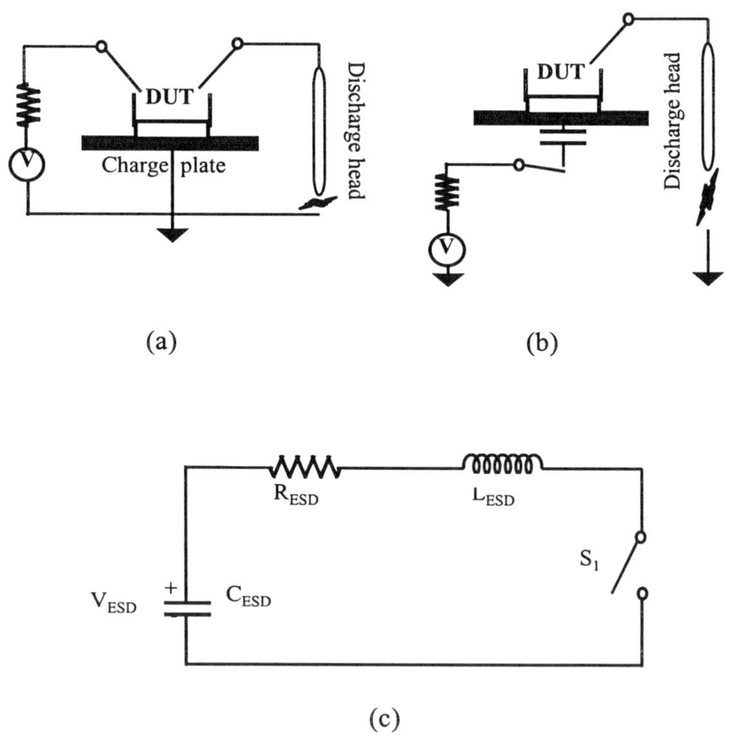

Figure 2.7 A CDM ESD test model. (a) Direct charging method, (b) Inductive charging method, (c) a simplified 2nd-order circuit model.

A CDM ESD simulator must be able to supply an ESD pulse waveform fitting with a standardized curve as illustrated in Figure 2.8 (solid line) and specified by Table 2.7. The specific means to realize CDM pulse generation is of course flexible as long as a pulse produced meets all the specifications. However, to build a meaningful and reliable CDM tester is extremely hard due to its sensitivity to parasitic effects in discharge paths. For example, assuming a very low L_{ESD} = 50nH can be realized, the ESD pulse produced by a second-order RLC network is still quite different from the standard one as shown in Figure 2.8 (dashed line). This is the main reason why there are no trustable CDM testers available for meaningful ESD tests yet, although the existence of CDM standards. Well, I heard you – *why should an IC designer care about all these nonsense?!* I feel exact the same as a circuit designer myself. However, a designer ought to have some know-how about the means others use to judge your work. This is the same reason why a quality IC designer always knows a great deal about process and device physics, in addition to enjoy the plots generated by SPICE on your screen. So, want to learn more? Please check with the Appendix A.

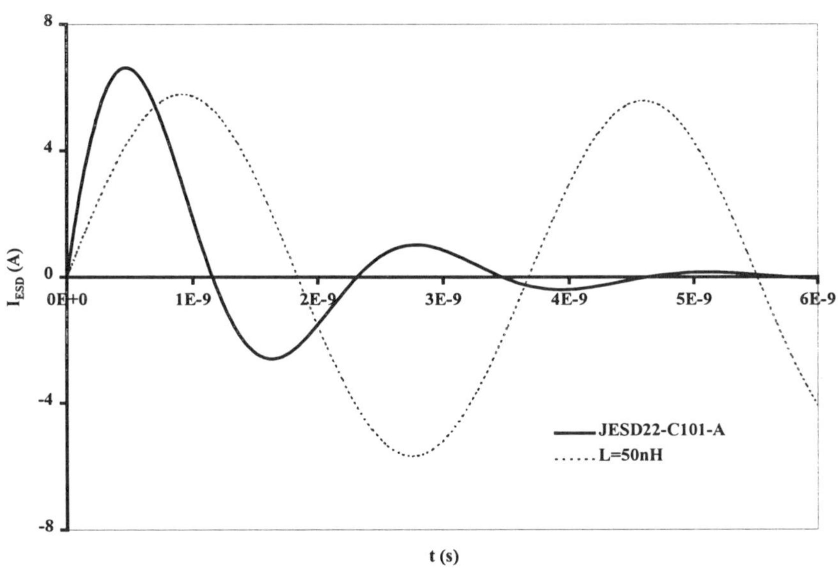

Figure 2.8 A standard CDM ESD pulse waveform specified by JESD22-C101-A (solid line). Parasitic effects (dashed line for L_{ESD} = 50nH) in discharge head have strong impact on the pulse generation. C_{ESD} = 6.8pF and V_{ESD} = 500V.

Table 2.7 Short-circuit CDM ESD waveform specifications per JESD22-C101-A for $C_{ESD} = 6.8pF$.

V_{ESD} (V)	Peak current I_p (A)	Undershoot I_{osc} (A)	Full width at half height (ns)	t_s (ns)
500	5.75 ± 15%	<50% of I_p	1.0 ± 0.5	<0.4

2.5. TLP MODEL

One common feature, disadvantageous, shared by all existing ESD test models is that the ESD test methods developed upon them are destructive. Such ESD zapping tests, to be discussed in Section 2.7, provide results about ESD failure threshold of an ESDS component. However, they offer no insights into the possible failure mechanisms, which are critical to ESD protection circuit design. Such information can be obtained by use of a transmission-line-pulsing (TLP) technique. Since its introduction into ESD protection design [20], the TLP technique has been increasingly accepted in practical ESD protection circuit design. The principle for TLP test is that a piece of transmission line cable is used to produce a stable square waveform, easier to control compared to that generated based upon other ESD test standards, to stress an ESDS DUT device. Many TLP testing systems with different flavour exist, among them a constant-impedance TLP set-up is illustrated in Figure 2.9a [21]. A piece of transmission line cable is pre-charged to a specific voltage level. It then discharges through a constant matching resistance, provided by another 50Ω-transmission line cable, into the ESDS DUT device. Instantaneous current and voltage (I-V) data are obtained by probing the current and voltage of the DUT, often using oscilloscopes. The DC leakage current under the operation power supply is measured after each ESD stress. Figure 2.9b shows the equivalent circuit for a TLP system, which is similar to those of other ESD test models, however, with a square waveform source replacing a charging capacitor. By incrementally stepping up the TLP pulse height, one can obtain an instantaneous I-V curve of the DUT under stressing and determine the possible ESD failure level by looking at the second breakdown point and DC leakage avalanche threshold. Figures 2.10 and 2.11 show example TLP test curves. An attenuator is used to absorb reflections from the DUT device. Critical parameters can be readily controlled. For example, the square voltage pulse height is $V_{TLP} = V_C/2$, where V_C is charging voltage; the desired current level is $I = (V_{TLP} - V_{DUT})/R$; the pulse duration is set by transmission length as t = 2(L/c), with c being the propagation velocity; and the desired pulse rise time is stabilized by the 50Ω constant impedance. TLP

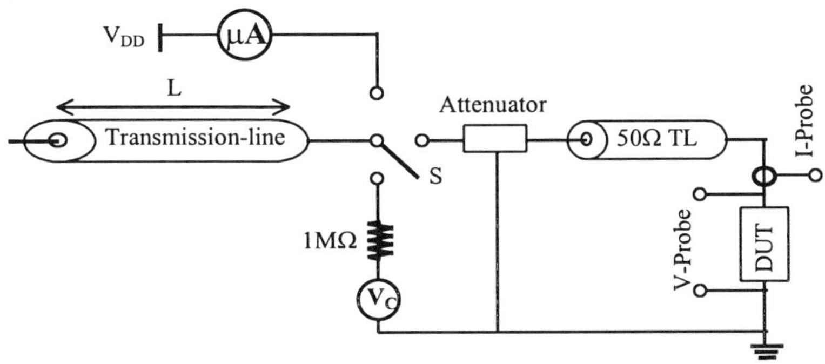

(a)

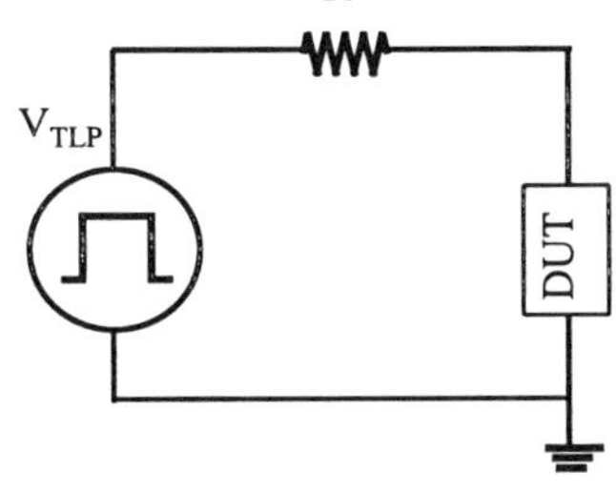

(b)

Figure 2.9 A typical TLP test set-up: (a) a constant-impedance TLP set-up. (b) an equivalent circuit for TLP tester.

test is quasi non-destructive in nature because of its well-controlled short duration. The instantaneous I-V curve and leakage current information obtained are essential for a designer to debug and optimise a design, whose applications will be discussed in case study examples in Chapter 8. One key point in using TLP testing is to correlating its results to those obtained using other ESD test models, i.e., HBM and CDM, for which controversial results

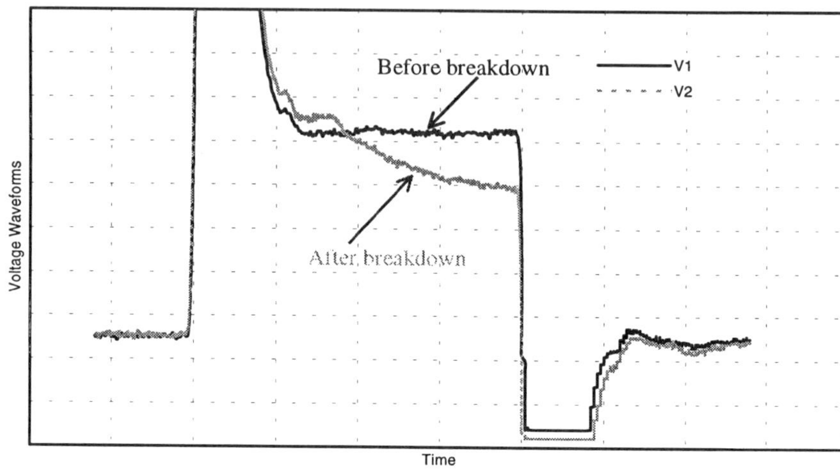

(a)

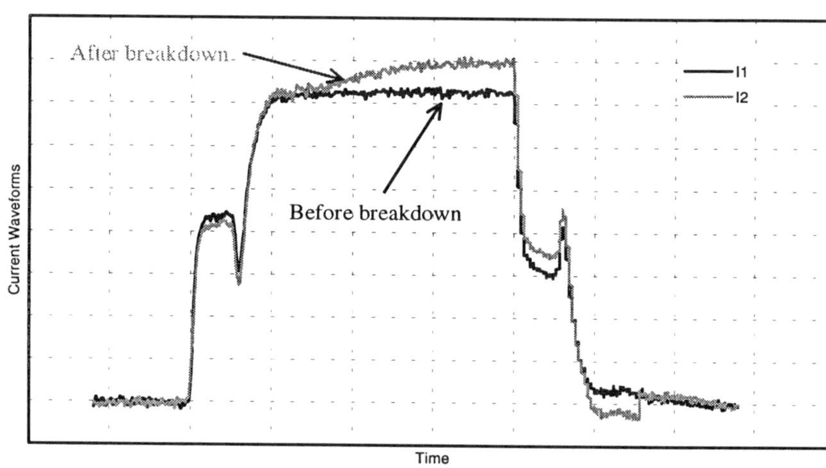

(b)

Figure 2.10 Sample TLP output voltage (a) and current (b) waveforms taken before (dark) and after (grey) second breakdown show voltage drop and current increase upon breakdown. (Courtesy of Barth Electronics, Inc. using Model 4002 TLP Tester)

were reported [22, 23]. Conceptually, such kind correlations can be ensured by setting TLP parameters, such as pulse duration and rise time, according to the test model concerned. For example, a duration of 100ns and a rise time of 10ns are typically used to match with HBM model. Though it is reported that good correlation factors were observed between TLP and HBM tests, one ought to be cautious that TLP set-up parameters play crucial roles in TLP tests. Because of the insights provided by TLP tests, it seems to be an imperative task to develop a standard TLP test model, which is currently under active investigation [21].

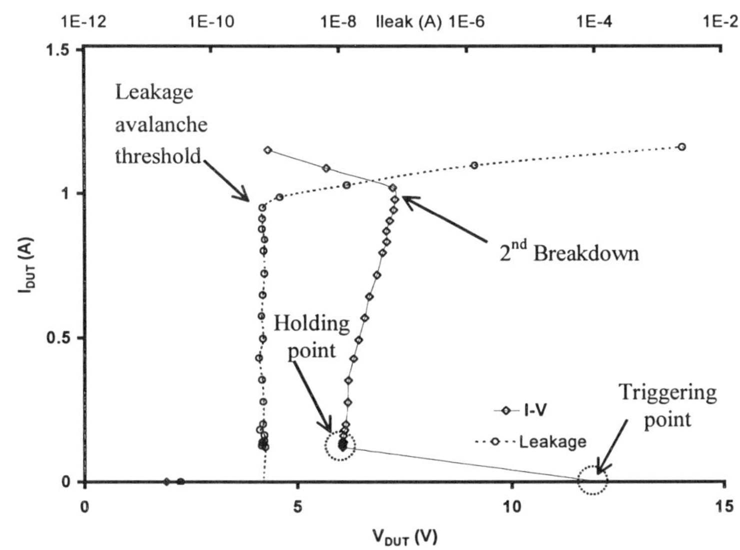

Figure 2.11 Example TLP data curves for an NMOS ESD structure: its damage current level is about 1A, correlated to HBM ESDV level of 1.5KV.

2.6. OTHER MODELS

In addition to the above-discussed ESD test models, many other specific test models were introduced, among them, an IEC model [24] might be of interest to IC designers. Although the IEC test model was initially developed for equipment ESD tests, many IC vendors begin to zap and specify products using IEC model to showcase the ESD toughness of the devices. The motivations behind include the similar human body origin between IEC and HBM, the sub-1ns rise time, the extremely high peak currents, and marketing strategy. As shown in Figure 2.12, the equivalent model circuit for IEC is similar to others, however with different charge capacitance, C_{ESD} = 150pF, and discharge resistance, $R_{ESD} = 330\Omega$. Figure 2.13 illustrates the

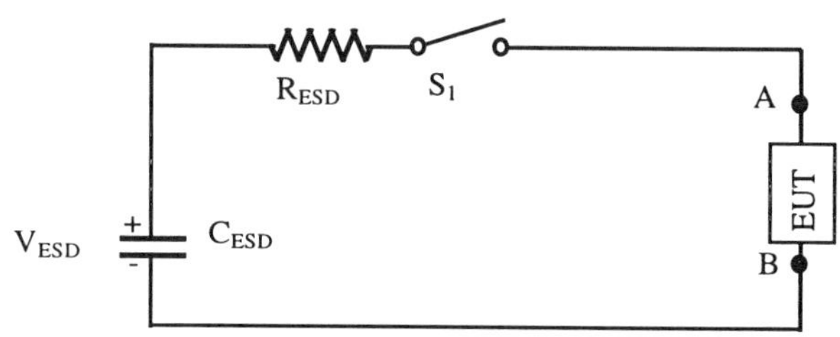

Figure 2.12 An equivalent circuit for IEC ESD test model.

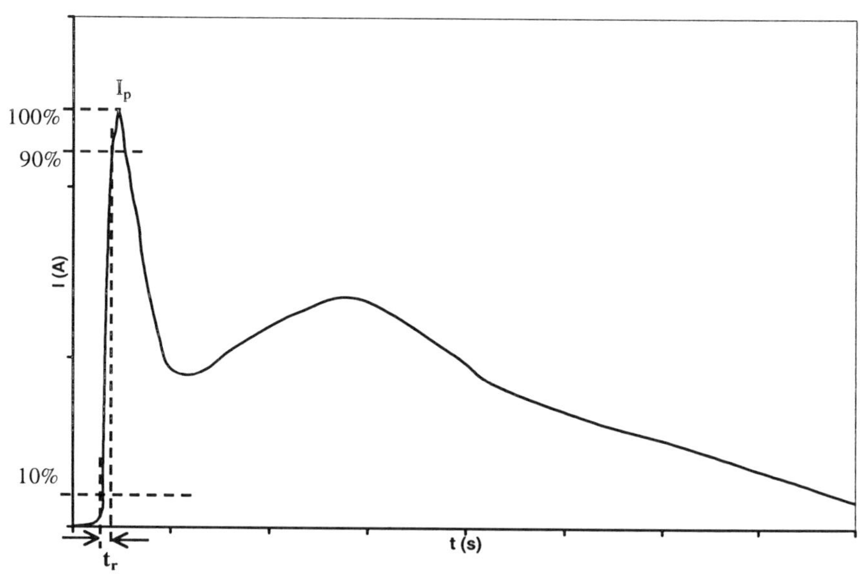

Figure 2.13 A typical IEC ESD pulse waveform.

standardized IEC pulse waveform with Table 2.8 summarizing its typical specifications. It is certainly that many inductive and/or capacitive decorations will be needed for an IEC model circuit to produce an ideal zap waveform. Because of its very short rise time, t_r = 0.7-1ns, and extremely high current peaks of up to several amperes, many suggests that complying to IEL ESD tests demonstrate very robust ESD immunity of components.

IEC test model suggests two discharge methods, e.g., air-gap discharge method that closely emulate a real world ESD event, but very dependent upon the moving speed of the discharge head towards a DUT and the gap in between; and contact discharge method whose results are more reproducible. More details are available in Appendix A. IEC test model has been included in component ESD testing systems by several equipment developers [25].

Table 2.8 Short-circuit IEC ESD waveform specifications per IEC 1000-4-2.

V_{ESD} (V)	Peak current I_p (A) ±10%	Current at 30ns (A)	Duration (ns)	t_s (ns)
1000	7.5	4	~ 80	0.7-1.0
4000	30	16	~ 80	0.7-1.0

It is worthy note that all these ESD test models basically share the same simplified model circuit as shown in Figure 2.3, however have different element values varying as system parasitic parameters. Table 2.9 gives a brief picture for the parametric differences. It is up to testing system developers to find out the "how" in generating standardized ESD pulses.

Table 2.9 A parametric matrix for various ESD test models.

Models	C_{ESD} (pF)	R_{ESD} (Ω)	Typical L_{ESD} (μH)
HBM	100	1500	7.5
MM	200	0*	0^+
CDM	6.8	0*	0^+
IEC	150	330	-

* Typically a few to a few tens ohms in discharge paths.

\+ Typically 1-2 μH in discharge paths.

2.7. ESD ZAPPING TESTS

ESD testing is generally referred to as ESD zapping test, which is used to determine the up most ESD sustainability level, e.g., ESDV, of an ESDS component. ESD zapping systems are constructed in compliance with selected ESD test standards to ensure supplying *ideal* ESD pulses, which are

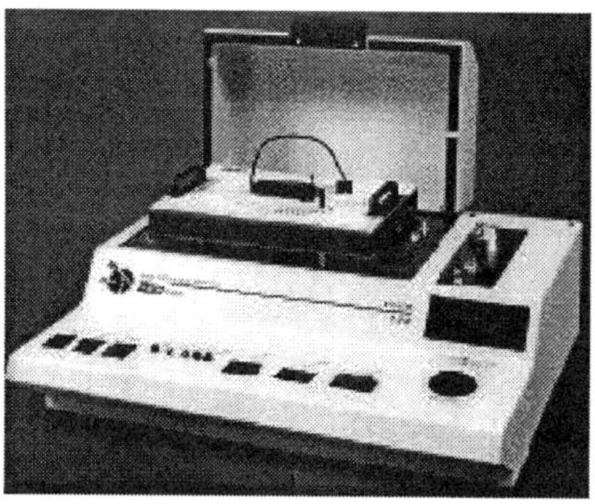

(a) System 700 by Oryx
Desktop, HBM, MM & IEC.

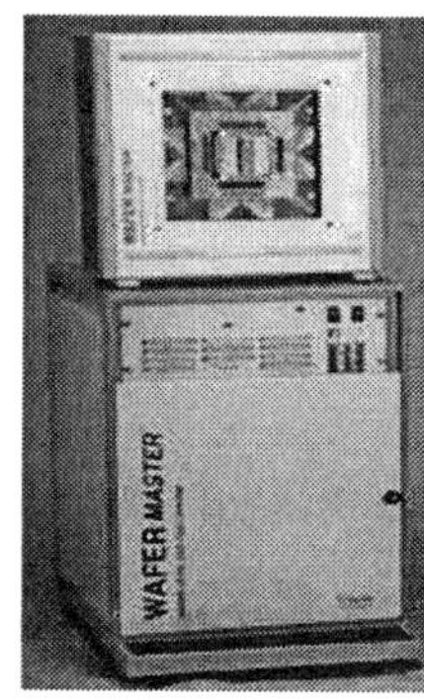

(b) WaferMaster by KeyTek
Wafer level, HBM & MM.

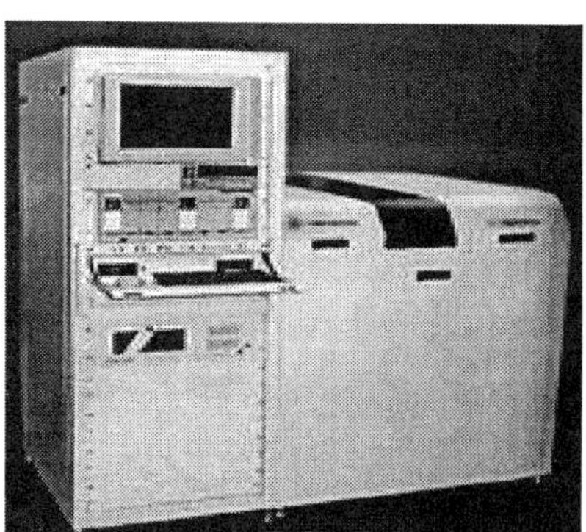

(c) System 11000EX by Oryx
HBM, MM, TLP & Latch-up.

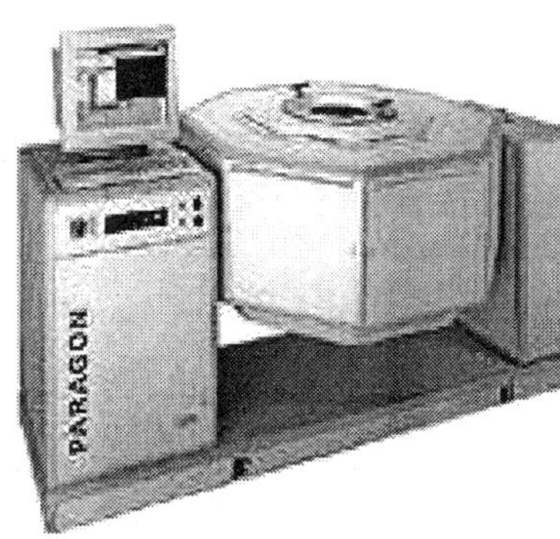

(d) PARAGON by KeyTek
HBM, MM & Latch-up.

Figure 2.14 Photos of some commercial ESD testing systems. (Courtesy of Oryx Instruments Corp. and Thermo KeyTek)

used to stress ESDS components. The goal of ESD zapping tests is to classify ESD robustness of IC parts, instead of providing insights into the failure mechanisms, because it only reports the pass or failure of a DUT under a given ESD pulsing level. Many commercial ESD testing systems are available in the market ranging from manual desktop model to very sophisticated systems for volume manufacturers, with Figure 2.14 showing a few of them. Figure 2.15 shows a photo of a TLP test system. A brief summary of commercial ESD testing systems is given in Appendix B.

Figure 2.15 A Model 4002 TLP Testing System (Courtesy of Barth Electronics, Inc.).

2.8. SUMMARY

ESD phenomena occur as static electricity transfers between two objects of different electrostatic potentials. The resulting ESD current and voltage transients can cause damages to IC parts. ESD events relevant to ICs are depicted by ESD models, such as, HBM, MM, CDM, TLP and IEC, corresponding to their origins. These ESD test standards are the bases for developing ESD testing systems as well as for ESD protection simulation and design. The discharges in these ESD test models can all be simplified to a second-order RLC network with different parametric values. Critical ESD discharge parameters must conform to the ideal discharge waveforms defined by various ESD standards. ESD robustness of ESDS IC parts are classified by performing ESD zapping measurements using ESD testing systems developed upon different ESD test standards. Available ESD test standards define *what* an ESD discharge waveform shall look like without specifying *how* to produce it. Detailed ESD testing procedures are provided in various ESD test standards targeting for different applications. While higher-order analytical models can be used to better fit standardized ESD discharge waveforms, real ESD pulses produced by an ESD tester is very sensitive to its parasitic parameters as well as that of DUT devices, making reproducible testing and cross-comparison of reported data a difficult task. Up to date, the most reliable, hence widely accepted, ESD test model is still the HBM model, which is used to specify almost all IC parts.

REFERENCES

[1] Dangelmayer, G. T., *ESD Program Management*, Boston, Kluwer Academic Publishers, 1999.

[2] Moss, R., "Caution – Electrostatic Discharge at Work", *IEEE Trans. Comp. Hyb. And Man., CHMT-5*, 1982, pp. 512-515.

[3] *MIL-STD-883E, Method 3015.7*, Electrostatic Discharge Sensitivity Classification, Dept. of Defence, Test Method Standard, Microcircuits, 1989.

[4] Van Roozendaal, L., Amerasekera, A., Bos, P., Baelde, W., Bontekoe, F., Kersten, P., Korma, E., Rommers, P., Krys, P., Weber, U., and Ashby, P., "Standard ESD Testing of Integrated Circuits", *Proc. EOS/ESD Symposium*, 1990, pp. 119 – 130.

[5] Amerasekera, A. and Verweij, J., "ESD in Integrated Circuits", *Quality and Reliability International, Vol. 8*, 1992, pp. 259 – 272.

[6] Verhaege, K., Roussel, P., Groeseneken, G., Maes, H., Gieser, H., Russ, C., Egger, P., Guggenmos, X., and Kuper, F., "Analysis of HBM ESD Testers and Specifications Using a 4th Order Lumped Element Model", *Proc. EOS/ESD Symposium*, 1993, pp. 129 – 137.

[7] *ESD STM5.1-1998 Revised*, Electrostatic Discharge Sensitivity Testing – Human Body Model Component Level, the ESD Association, 1998.

[8] *JEDEC JESD22-A114-B*, Electrostatic Discharge Sensitivity Testing Human Body Model, the Electronics Industries Alliance, June 2000.

[9] *AEC – Q100-002 Rev-C*, Human Body Model Electrostatic Discharge Test, the Automotive Electronics Council, October 1998.

[10] *EIAJ-Standard, IC-121-1988*, Testing Method 20: Electrostatic Destructive Test, Electrostatic Discharge, 1988.

[11] *ESD STM5.2-1999 Revised*, Electrostatic Discharge Sensitivity Testing – Machine Model Component Level, the ESD Association, 1999.

[12] *JEDEC JESD22-A115-A*, Electrostatic Discharge Sensitivity Testing Machine Model, the Electronics Industries Alliance, October 1997.

[13] *AEC – Q100-003 Rev-E*, Machine Model Electrostatic Discharge Test, the Automotive Electronics Council, January 2001.

[14] Speakman, T., "A Model for Failure of Bipolar Silicon Integrated Circuits Subjected to Electrostatic Discharge", *Proc. IEEE Int'l Rel. Phys. Symp.*, 1974, pp. 60-69.

[15] Bossard, P., Chemelli, R. and Unger, B., "ESD Damage from Triboelectrically Charged IC Pins", *Proc. EOS/ESD Symp.*, 1980, pp. 17-22.

[16] Chaine, M., Verhaege, K., Avery, L., Kelly, M., Gieser, H., Bock, K., Henry, L., Meuse, T., Brodbeck, T., and Barth, J., "Investigation into Socketed CDM Tester Parasitics", *Proc. EOS/ESD Symp.*, 1998, pp. 301-310.

[17] *ESD STM5.3.1-1999 Revised*, Electrostatic Discharge Sensitivity Testing – Charged Device Model Component Level, the ESD Association, 1999.

[18] *JEDEC JESD22-C101-A*, Field-Induced Charged-Device Model Test Method for Electrostatic Discharge-Withstand Thresholds of Microelectronic Components, the Electronics Industries Alliance, June 2000.

[19] *AEC – Q100-011 Rev-A*, Charged Device Model Electrostatic Discharge Test, the Automotive Electronics Council, January 2001.

[20] Maloney, T. and Khurana, N., "Transmission Line Pulsing Technique for Circuit Modelling of ESD Phenomena", *Proc. EOS/ESD Symp.*, 1985, pp. 49-54.

[21] Barth, J., Verhaege, K., Henry, L., and Richner, J., "TLP Calibration, Correlation, Standards, and New Techniques", *Proc. EOS/ESD Symp.*, 2000, pp.85-96.

[22] Stadler, W., Guggenmous, X., Egger, P., Gieser, H. and Mussshoff, C., "Does the ESD Failure Current Obtained by Transmission-Line Pulsing Always Correlate to Human Body Model Tests?" *Proc. EOS/ESD Symp.*, 1997, pp.366-372.

[23] Amerasekera, A., Roozendaal, L., Abderhalden, J., Bruines, J., and Sevat, L., "An Analysis of Low Voltage ESD Damage in Advanced CMOS Processes", *Proc. EOS/ESD Symp.*, 1990, pp. 143-150.

[24] *IEC 1000-4-2*, "Electromagnetic Compatibility, Part 4: Testing and Measurement Techniques, Section 2: Electrostatic Discharge Immunity Test", The International Electrotechnical Commission, January 1995.

[25] Commercial ESD testers including IEC test model: *System 700* by Oryx Instruments Corp. and *ZAPMASTER 7/1-7/4* by Thermo KeyTek.

Chapter 3

ESD PROTECTION DEVICE SOLUTIONS

This chapter discusses general ESD protection principles, basic devices used as ESD protection structures, and the underlying device physics. Understanding of these ESD protection device fundamentals are essential to successful design of ESD protection circuits, which will be discussed in Chapter 4. This is particularly true as IC technologies become more complex and IC chips get more sophisticated.

3.1. ON-CHIP ESD PROTECTION MECHANISMS

The two main ESD failure mechanisms are associated with thermal damages and dielectric failures. The root cause to thermal damage is the very high transient current generated by an ESD pulse. Since semiconductors, e.g., silicon, are normally poor thermal conductors, the heat generated by large ESD currents cause temperature increase in both silicon and metal interconnect materials. The situation is even worse because heat generators are usually located near the surface covered by very poor thermal conductive silicon dioxide films. Because of the transient nature of ESD events, the heat generated has little chance to be dissipated. Consequently, ESD damages occur as highly localized thermal defects in both silicon and metals. On the other hand, dielectric failures are caused by high electric field induced dielectric breakdown, often occurring in MOSFET gate oxide layers. For example, given typical SiO_2 dielectric strength of 8~10 MV/cm [1], a gate oxide of 35 Å in a 0.18μm CMOS technology could be damaged by a bias of less tan 4V.

Having known the root causes to ESD damages, one ought to deal with both large currents and high voltages in ESD protection design. Therefore, the principle of on-chip ESD protection solutions is two-fold: to safely discharge ESD currents via a low-impedance shunting path and to clamp a bond pad voltage to a sufficiently low level. There are two general means to realize the above ESD protection concepts, as illustrated in Figure 3.1. The first option is to use a protection device with a simple turn-on I-V characteristic as shown in Figure 3.1a, where the protection unit is turned on at a threshold point (V_{t1}, I_{t1}) with t_1 being the triggering time and forms a low-impedance shunt channel to discharge ESD transients. The turn-on voltage V_{t1} should be sufficiently low for voltage clamping, however, shall be greater than operation voltage of the IC chip, i.e., V_{DD}, to avoid accidental turn-on under normal operation. The current handling capability, reflecting the ESDV level, is only limited by heat generation due to series resistance in the shunting channel. This type of ESD protection devices can be readily included into circuit simulation. The second solution is based upon a snapback I-V characteristic as depicted in Figure 3.1b, where a protection element is turned on at a triggering point (V_{t1}, I_{t1}, t_1), then driven into a snapback region with low holding (V_h, I_h) to create a low-impedance discharge path. The trigger voltage V_{t1} is designed according to the IC chip. The holding voltage V_h shall ensure proper voltage clamping and the holding current I_h is selected with latch-up in consideration. A deep snapback I-V is therefore preferable for better ESD protection. The ESD protection performance level, i.e., ESDV value, is typically represented by the second

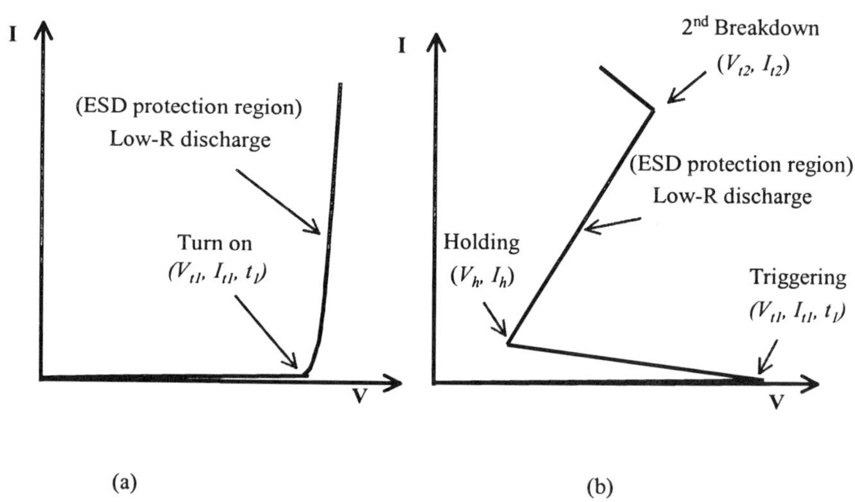

Figure 3.1 Typical I-V characteristics used for ESD protection design: (a) simple turn-on; (b) snapback.

breakdown (or, thermal breakdown) current I_{t2}. While early-day ESD protection usually resorts to the simple turn-on type protection devices, such as forward junction diodes; a snapback type protection is more efficient and gaining popularity nowadays. However, a main disadvantage of the snapback type design is that it cannot be included in circuit simulation. It is worth to point it out that critical aspect in rational ESD protection design shall be to properly and accurately select the triggering point (V_{t1}, I_{t1}, t_1), holding point (V_h, I_h) and second breakdown threshold (V_{t2}, I_{t2}), which are key parameters in ESD protection circuit design. Most ESD protection solutions are developed based upon the above two principles. Typical ESD protection devices, being the building bricks for all ESD protection networks, are discussed in the following sections in details, with more sophisticated ESD protection circuits to be discussed in Chapter 4.

3.2. DIODE AS ESD PROTECTION ELEMENT

Junction diodes are the dinosaurs of ESD protection. However, owing to its simplicity and efficiency, a diode ESD protection is still widely used on today's IC chips. It is noteworthy that a simple solution is always the best solution in any design, including ESD protection, as long as adequate protection is achieved. Artistic work might push the limit, however, any bells and whistles will not come for free. The ESD-induced parasitic effects must be taken into consideration as to be discussed in Chapter 9.

Diodes can be used as ESD protection devices in either forward-biasing fashion or reverse-biasing format, typically as Zener diodes. Either way, the simple turn-on I-V characteristic is in charge. It is certainly worth to revisit some diode physics.

3.2.1. Diode Device Physics

Figure 3.2 shows a typical diode I-V characteristic. In forward-biasing condition, a diode starts to conduct significant current after turn-on, typically at $V_{ON} \sim 0.65V$ for silicon diode. The I-V characteristic of an ideal diode, expressed in total values, follows the Shockley equation [2, 3],

$$i_D = I_S(e^{\frac{v_D}{nV_T}} - 1) \tag{3.1}$$

and

$$I_S \equiv Aqn_i(\frac{D_P}{L_P N_D} - \frac{D_N}{L_N N_A}) \tag{3.2}$$

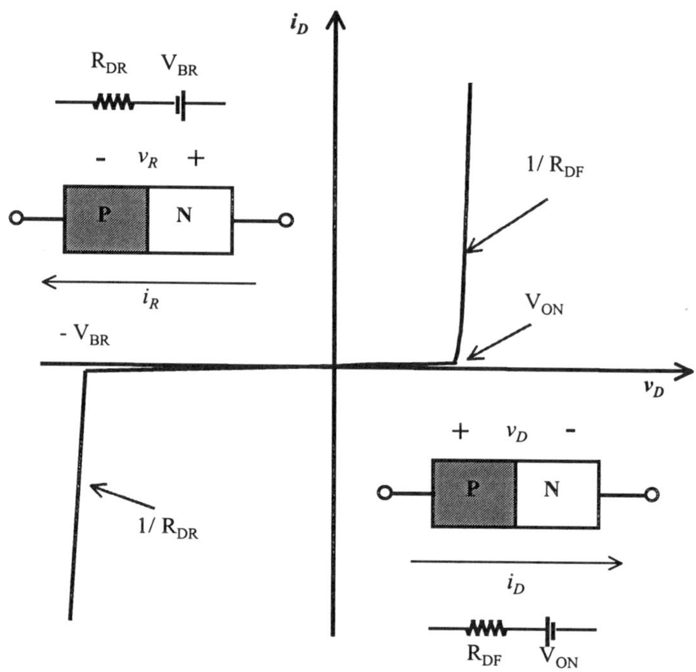

Figure 3.2 Typical diode I-V characteristics, including piece-wise-linear models.

where v_D and i_D are diode biasing voltage and current, n ≈ 1~2, thermal voltage $V_T \equiv kT/q$, k is Boltzmann constant, T is temperature, q is single electron charge, I_S is called saturation current, A is diode junction size, n_i is intrinsic carrier density, D_x, L_x and N_x are diffusion constants, diffusion lengths and impurity densities for donors and acceptors, respectively. The diode can be depicted by a piece-wise-linear model as shown in the inset of Figure 3.2 and follows the equation [4],

$$v_D = i_D R_{DF} + V_{ON} \tag{3.3}$$

where R_{DF} is diode forward on-resistance, which is an important parameter in ESD protection design. The temperature dependence, a critical factor in ESD operation, can be approximated by [3]

$$I_S \approx T^{(3+\gamma/2)} e^{-\frac{E_g}{kT}} \tag{3.4}$$

where E_g is bandgap of silicon and γ is a constant.

In reverse-biasing mode, only negligible reverse leakage current, i_R (= -i_D) ≈ I_S, should be observed until breakdown occurs at a breakdown voltage of v_D = - v_R = -V_{BR}. The corresponding piece-wise-linear model is give by the inset of Figure 3.2 and follows the equation [4]

$$v_R = i_R R_{DR} + V_{BR} \tag{3.5}$$

where R_{DR} is equivalent diode on-resistance in reverse mode. Typical diode reverse breakdown is caused by avalanche multiplication, or impact ionisation, in depletion region governed by [3, 5]

$$V_{BR} = \frac{\varepsilon_s E_{max}^2}{2qN_B} \tag{3.6}$$

and

$$M = \frac{1}{1-(\frac{v_R}{V_{BR}})^n} \tag{3.7}$$

where ε_s is semiconductor permittivity, E_{max} is maximum electric field in the depletion region, N_B is impurity density of the lighter region in a single-sided abrupt PN junction, and n is a constant.

3.2.2. Diode in ESD Protection Operation

As an ESD protection device, a diode can be operated in both forward and reverse biasing regions. In forward-biasing mode, the previously discussed device physics generally governs. The main difference is that the diode is operating in high-current mode, which is actually the case for all devices under ESD protection. Under high-current condition, both drift and diffusion current components must be considered. Further, the potential drop across the intrinsic PN junction becomes insignificant compared to the ohmic drop over the diode series resistance, r_S, under high injection (~10^4 A/cm^2). Hence, the diode I-V characteristic in ESD events can be approximated by [3]

$$i_D \propto e^{\frac{v_D}{2V_T}} \tag{3.8}$$

Typical diode ESD protection scheme is illustrated in Figure 3.3 for both forward and reverse biasing modes. Because of its low forward turn-on voltage, single forward diode normally cannot be used as ESD protection device for power supply of V_{DD} > 5V. A diode string can be used instead with the number of diodes in the chain depending upon IC chips. However, the diode string connection causes Darlington effect in IC fashion, which will be discussed in Chapter 4. In contrast to forward-biasing mode, a reverse connected diode is the simplest ESD protection solution, where its

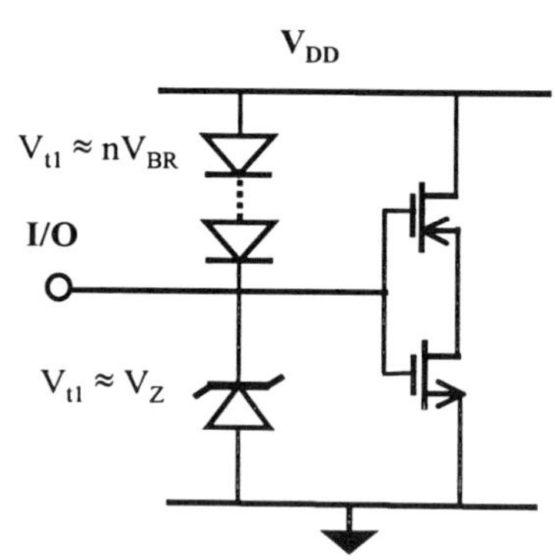

Figure 3.3 Typical diode ESD protection schemes can use both forward and reverse diode connection.

triggering voltage, V_{t1}, is determined by reverse breakdown voltage, V_{BR}. In practices, a rich variety of P/N layers in an IC process technology can be selected to obtain a desired V_{t1} fitting a specific IC chip. Typically, a Zener diode is used as an ESD protection device with $V_{t1} \approx V_Z$, the Zener diode turn-on voltage. Careful design and layout for the diodes are crucial to minimizing diode series resistance, r_S, for higher current handling capacity. One extra major advantage in using diode ESD protection is that these diodes can be included in normal circuit simulation.

3.2.3. Diode Parasitic Modelling

As an IC designer, one ought to think of chip-level interactions between ESD protection devices and circuits being protected. As always, nothing comes for free, even in the case of using the simplest diodes. Such full-scale consideration will be discussed in Chapter 9. Here, typical parasitic models for ESD diodes are introduced. ESD-induced parasitic capacitance, C_{ESD}, is a real design problem that not only leads to excess RC delay, but also causes external noise injection due to C_{ESD} coupling. A small-signal model including C_{ESD} is given in Figure 3.4 for an ESD protection diode, where $C_{ESD} = C_D$ includes both depletion and diffusion capacitances of a PN junction diode. A noise model is also depicted in Figure 3.4, where r_d and r_S

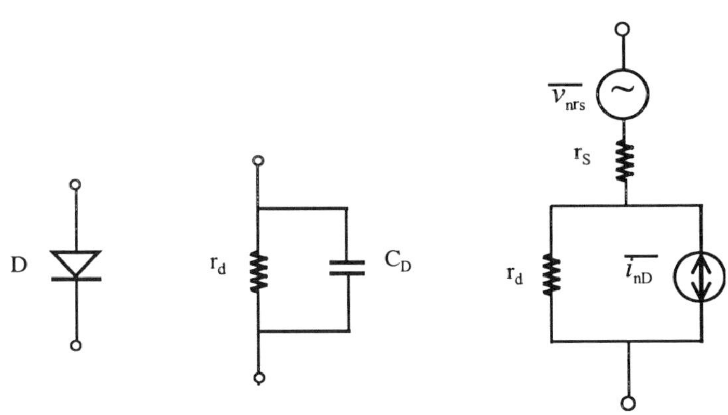

Figure 3.4 Parasitic capacitance and noise models for a diode.

are diode dynamic and series resistances, $\bar{v}_{nr_S}$ is r_S–induced thermal noise, and $\bar{i}_{nD}$ is intrinsic diode noises consisting of shot and flicker noises. The noise power spectral densities are given by [6, 7]

$$\bar{v}^2_{nr_S} = 4kTr_S\Delta f \tag{3.8}$$

and

$$\bar{i}^2_{nD} = 2qi_D\Delta f + K\frac{i_D^a}{f}\Delta f \tag{3.9}$$

where Δf is the bandwidth of interest, and K and a are process and device related coefficients. These parasitic models can be used in evaluating the influences of ESD protection structures on circuit performance.

3.3. BJT AS ESD PROTECTION ELEMENT

Advanced ESD protection relies on active discharging using bipolar junction transistors (BJT), MOS field-effect transistors (MOSFET), silicon-controlled rectifiers (SCR), and their sub-circuits. A BJT is actually the underlying building bricks of these advanced ESD protection networks. Hence, it is important to apprehend BJT operation.

3.3.1. BJT Device Physics

Assuming an ideal NPN BJT transistor is connected in common emitter (CE) mode as shown in Figure 3.5, depending upon the junction biasing voltages, a BJT can work in three different regions, i.e., cut-off if both junctions are reverse-biased ($V_{BE} <0$ & $V_{BC} <0$), active if emitter junction is forward-biased while collector junction is reverse-biased ($V_{BE} >0$ & $V_{BC} <0$) and saturation if both junctions are forward-biased ($V_{BE} >0$ & $V_{BC} >0$). Looking into active mode, current amplification occurs because minority carriers, i.e., electrons, are injected from emitter into base, where most of them are swept into collector, with a small fraction of the minorities are consumed in base due to recombination with holes. Similarly, holes in base are injected into emitter. External base current flows into base to continuously supply holes, therefore maintain the BJT operation. Significant current amplification from base to collector is observed because a small variation in base current leads to large change in collector current. The terminal I-V characteristics can be approximated by the following equations [3, 4, & 7],

$$i_D \approx I_S e^{\frac{v_{BE}}{V_T}} \tag{3.10}$$

$$I_S \equiv \frac{A_E q D_n n_i^2}{N_A W_B} \tag{3.11}$$

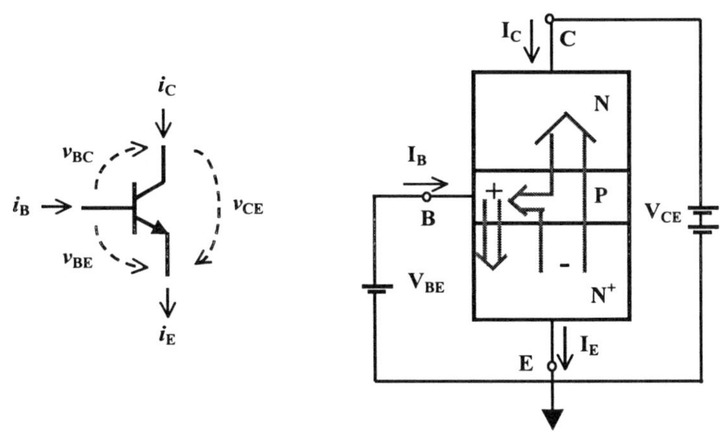

Figure 3.5 A CE connection for an NPN BJT transistor.

$$i_B \approx I_S \left(\frac{D_p}{D_n}\frac{N_A}{N_D}\frac{W_B}{L_p} + \frac{1}{2}\frac{W_B^2}{D_n \tau_b}\right) e^{\frac{v_{BE}}{V_T}} \tag{3.12}$$

and

$$i_E = i_B + i_C \tag{3.13}$$

where i_E, i_B, and i_C are total emitter, base and collector currents, v_{BE} is total emitter junction bias, A_E is emitter junction size, W_B is base width, N_A and N_D are acceptor and donor impurity densities, D_x is diffusion coefficient, L_x is diffusion length, and τ_b is minority carrier lifetime in the base. The CE (common emitter) and CB (common-base) current gains, i.e., β_F & α_F for DC forward mode, are therefore defined as

$$\beta_F \equiv \frac{I_C}{I_B} = \frac{1}{\frac{D_p}{D_n}\frac{N_A}{N_D}\frac{W_B}{L_p} + \frac{1}{2}\frac{W_B^2}{D_n \tau_b}} \tag{3.14}$$

and

$$\alpha_F \equiv \frac{I_C}{I_E} = \frac{\beta_F}{1+\beta_F} \tag{3.15}$$

where I_E, I_B & I_C are DC terminal currents and footnote F indicates forward operation.

Typical NPN BJT terminal I-V characteristics are depicted in Figure 3.6, where the incremental increase of collector current in active region comes from Early effect, with V_A being the Early voltage. Continuous increase of reverse bias V_{BC} will lead to avalanche breakdown (or punch-through breakdown if W_B is narrow enough) in collector junction and avalanche multiplication as depicted by

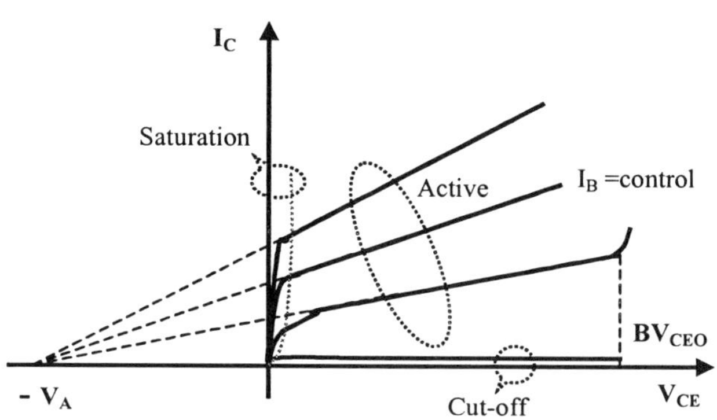

Figure 3.6 Typical CE terminal I-V characteristics for an NPN BJT.

$$M = \frac{1}{1-(\frac{v_{CB}}{BV_{CBO}})^n} \tag{3.16}$$

and

$$BV_{CEO} = BV_{CBO}(1-\alpha_F)^{\frac{1}{n}} \tag{3.17}$$

where BV_{CEO} and BV_{CBO} are open-base breakdown and open-emitter breakdown voltages, respectively. This avalanche multiplication effect plays a key role in BJT ESD protection operation.

A universal device model for a BJT transistor is the Ebers-Moll model as shown in Figure 3.7, with typical terminal I-V equations given by

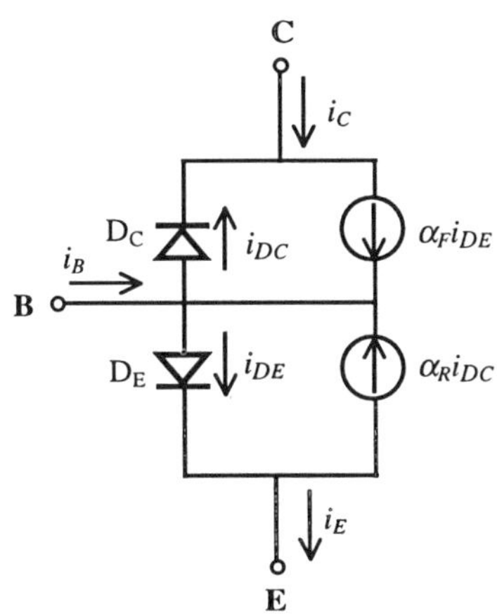

Figure 3.7 An Ebers-Moll model for an NPN BJT.

$$i_E = \frac{I_S}{\alpha_F}(e^{\frac{v_{BE}}{V_T}}-1) - I_S(e^{\frac{v_{BC}}{V_T}}-1) \tag{3.18}$$

$$i_C = I_S(e^{\frac{v_{BE}}{V_T}} - 1) - \frac{I_S}{\alpha_R}(e^{\frac{v_{BC}}{V_T}} - 1) \tag{3.19}$$

and

$$i_B = \frac{I_S}{\beta_F}(e^{\frac{v_{BE}}{V_T}} - 1) + \frac{I_S}{\beta_R}(e^{\frac{v_{BC}}{V_T}} - 1) \tag{3.20}$$

where footnotes F and R indicate forward and reverse operations, respectively. A simplified version can be readily obtained for active mode operation.

3.3.2. BJT in ESD Protection Operation

A BJT normally works in a snapback mode as an ESD protection device. Typical BJT ESD protection schemes are illustrated in Figure 3.8, where the

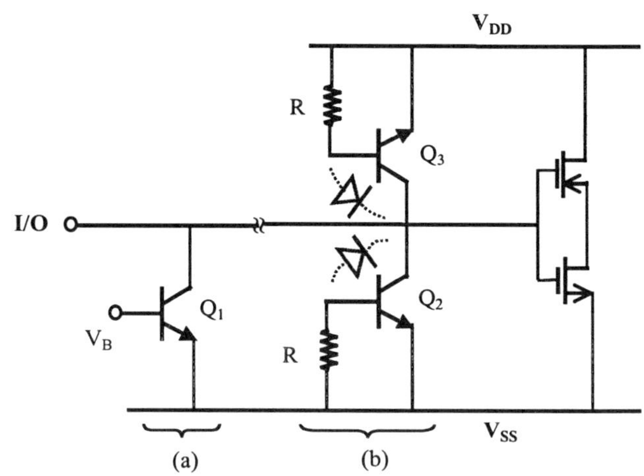

Figure 3.8 Basic BJT ESD protection schemes.

collector of a BJT is connected to an I/O pad. Take the connection case (a) in Figure 3.8 as a general example; BJT Q_1 is in off-state in normal operation because its collector and emitter junctions are reverse-biased. When an ESD pulse appears at I/O pad, the design ensures to raise the base potential, V_B, to turn on Q_1. An active discharge channel is then formed to shunt the ESD transient. A simple practical topology is shown in Figure 3.8 as connection case (b). As an ESD pulse comes to the I/O pad with respect to V_{SS}, the collector junction of BJT Q_2 is reverse-biased to its breakdown. After avalanche multiplication takes off at collector, hole current will be collected by V_{SS} through resistor R, which builds up the voltage across the emitter junction. As V_{BE} becomes greater than V_{ON} of emitter junction, Q_2 is turned on. Once the Q_2 turns on, emitter current i_E takes over the role of v_{CB} in maintaining the multiplication. V_C starts to decrease and Q_2 moves into snapback region as shown in Figure 3.9. A low-impedance channel is hence

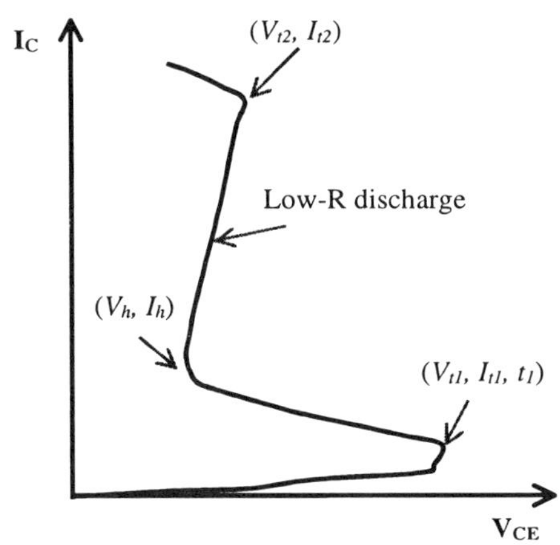

Figure 3.9 A snapback I-V characteristic for a BJT in ESD protection operation.

formed to safely discharge ESD current transients. In the mean time, the pad voltage is clamped to a low holding voltage level, V_h, presumably low enough to avoid dielectric rupture. If a negative ESD pulse appears at the I/O pad with reference to V_{SS}, the BE junction diode will be forward turned on to form a shunt path. Figure 3.10 shows a conceptual cross-section for a BJT ESD protection structure. In practical design, one shall expect a good deal of sweat in setting up the triggering and holding points properly, although it does sound fairly easy in talking: the V_{t1} is determined by BV_{CBO}; while the

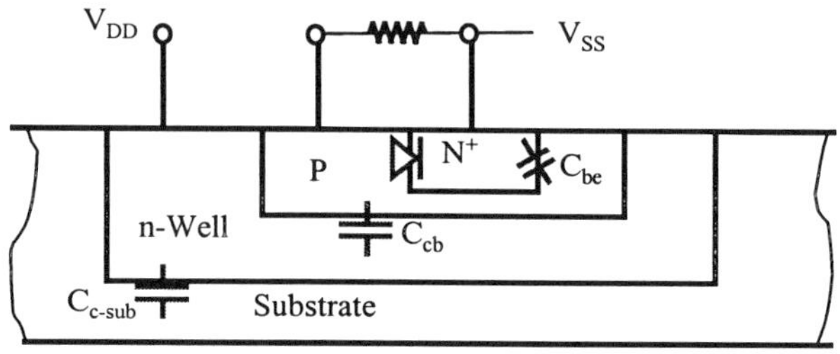

Figure 3.10 A cross-section for a BJT as an ESD protection device.

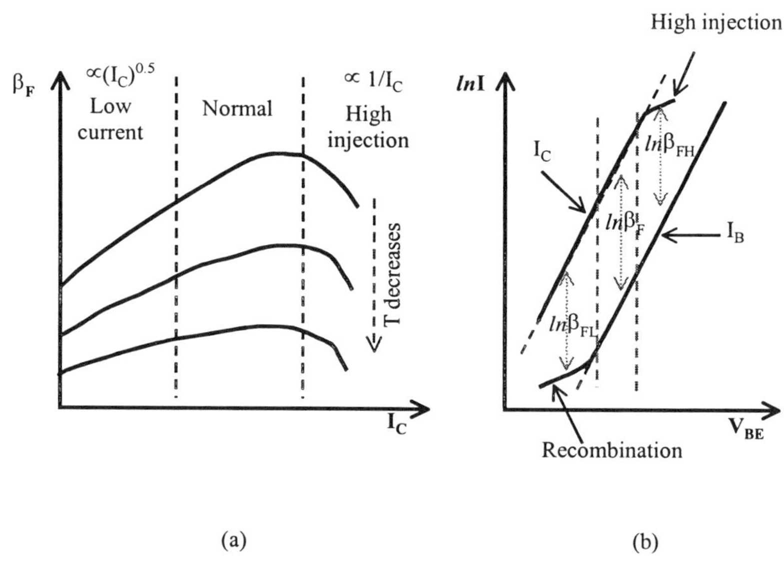

Figure 3.11 Typical BJT gain varies according to biasing. BJT ESD protection devices work in high-injection region.

snapback depth and V_h are directly related to impact ionisation, base width, BJT current gain, and resistance in the path. Since a BJT operates in high-current condition under ESD, it is noteworthy to look into high-injection

BJT effect as illustrated in Figure 3.11, where the nominal current gain remains good in a limited range of I_C beyond that β drops dramatically. At low-current region, decrease in β is due to extra base current from recombination in EB junction depletion region. In high-current region, the root causes are associated with series resistances and the high-current-injection effect [3, 8]. More sophisticated BJT-based ESD protection scheme will be discussed in Chapter 4.

3.3.3. BJT Parasitic Modelling

Parasitic effects of BJT transistors, as ESD protection units, should be included into IC chip design as well, although those BJT ESD devices are off in normal operation. However, relative importance of individual parasitic elements are certainly different under ESD condition compared to in normal circuit operation. Typical parasitic ESD-BJT capacitances, C_{ESD}, can be analysed based upon its cross-section, as shown in Figure 3.10. As usual, all junction capacitances should be considered. However, ESD-BJT remains in off-state normally, the base charging capacitance can be ignored. Clearly, different ESD protection topology shall result in different parasitic behaviours. Taking the BJT connection case (b) shown in Figure 3.8 as an example, its parasitic capacitance network can be modelled by that shown in Figure 3.12, where BC junction capacitance, C_{cb}, and collector-substrate capacitance, $C_{c\text{-}sub}$, dominate. The forward-biased BE junction capacitance, C_{be}; does not play much a role. Hence, the total ESD-induced parasitic capacitance for one BJT is $C_{ESD} \approx C_{cb}// C_{c\text{-}sub}$. The ESD-induced noise model

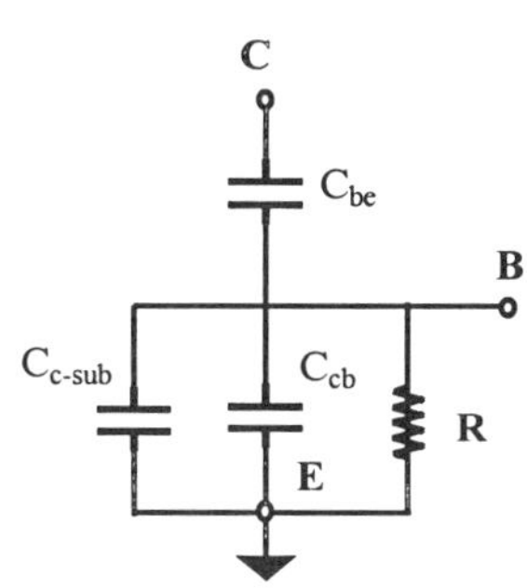

Figure 3.12 A parasitic C_{ESD} model for the BJT ESD protection network in Figure 3.8 (b).

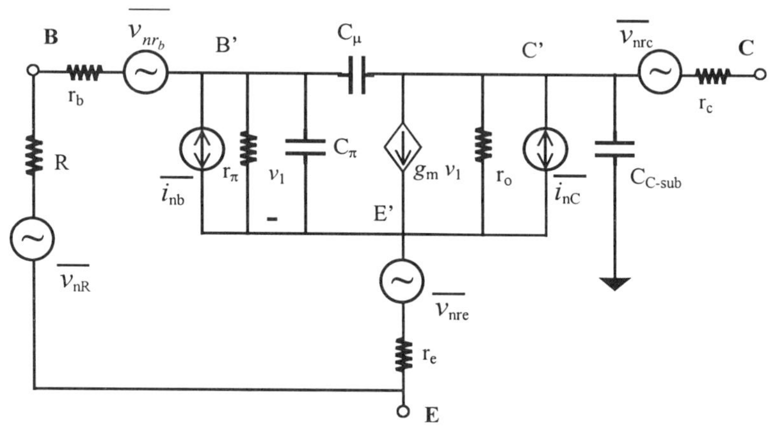

Figure 3.13 A noise model for the BJT ESD protection device in Figure 3.8 (b).

is modelled by the noise network shown in Figure 3.13, with formulas for each noise generators given by [4, 6]

$$\overline{i_{nC}^2} = 2qi_C\Delta f \tag{3.21}$$

$$\overline{i_{nb}^2} = 2qi_B\Delta f + K_1\frac{i_B^a}{f}\Delta f \tag{3.22}$$

$$\overline{v_{nr_b}^2} = 4kTr_b\Delta f \tag{3.23}$$

$$\overline{v_{nr_e}^2} = 4kTr_e\Delta f \tag{3.24}$$

$$\overline{v_{nr_c}^2} = 4kTr_c\Delta f \tag{3.25}$$

and

$$\overline{v_{nR}^2} = 4kTR\Delta f \tag{3.26}$$

where R is external resistance, r_b, r_e, and r_c are BJT series resistances, K_1 and a are device-specific constants, and Δf is the frequency range of interest. The noises considered included thermal noises (equations 3.23 to 3.26), shot noises (equation 3.21 and the first term in equation 3.22), and Flicker noise in BE junction (second term in equation 3.22). Depending upon the actually operation, other noise generators may be considered.

3.4. MOSFET AS ESD PROTECTION ELEMENT

MOS field effect transistor (MOSFET) is currently the most widely used ESD protection structures because of its active discharge mechanism and compatibility to CMOS technologies. However, design of MOS ESD protection for high ESD protection performance (>2KV) is not an easy task. As CMOS IC technologies advance, design of efficient and robust MOS ESD protection networks becomes extremely challenging. Understanding the underlying basics is very beneficial.

3.4.1. MOSFET Device Physics

A typical enhancement mode NMOSFET is illustrated in Figure 3.14.

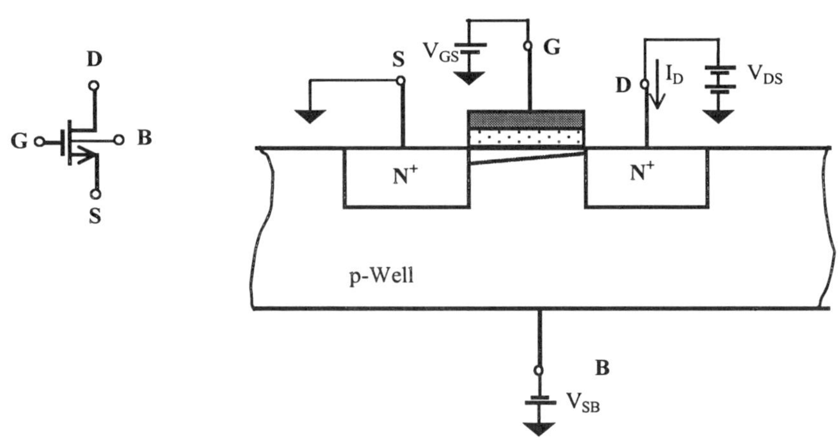

Figure 3.14 Cross-section and biasing condition for an enhancement mode NMOSFET.

Assuming a threshold voltage of V_{th}, the NMOSTFET operation can be divided into three regions depending upon its biasing conditions, i.e., cut-off region if $V_{GS} < V_{th}$; linear region if $V_{GS} > V_{th}$ and $V_{GD} = V_{GS} - V_{DS} > V_{th}$; and saturation region if $V_{GS} > V_{th}$ and $V_{GD} = V_{GS} - V_{DS} < V_{th}$. The transistor's terminal characteristics, including second-order effects, can be depicted by the following equations: [3, 4, 8]

threshold voltage,

$$V_{th} = V_{tho} + \frac{\sqrt{2qN_A\varepsilon_s}}{C_{ox}}(\sqrt{2\varphi_f + V_{SB}} - \sqrt{2\varphi_f}) \quad (3.27)$$

$$V_{tho} = \phi_{ms} + \frac{Q_b}{C_{ox}} - \frac{Q_{SS}}{C_{ox}} + 2\varphi_f \quad (3.28)$$

linear current,

$$i_D = \mu_n C_{ox} \frac{W}{L_{eff}}[(V_{GS} - V_{th})V_{DS} - \frac{1}{2}V_{DS}^2] \quad (3.29)$$

saturation current,

$$i_{Dsat} = \frac{1}{2}\mu_n C_{ox} \frac{W}{L_{eff}}(V_{GS} - V_{th})^2(1 + \lambda V_{DS}) \quad (3.20)$$

and sub-threshold current,

$$i_{Dsub} = K\mu_n C_{ox} \frac{W}{L_{eff}} e^{(V_{GS} - V_{th})/nV_T}(1 - e^{-mV_{DS}/nV_T}) \quad (3.21)$$

where V_{tho} is intrinsic threshold voltage defined by metal-semiconductor work function difference, ϕ_{ms}, surface depletion region space charges, Q_b, fixed charges in gate oxide, Q_{SS}, and strong inversion Fermi level, $2\varphi_f$; C_{ox} is unit gate oxide capacitance; μ_n is electron mobility; W is channel width; L_{eff} is effective channel length; $1/\lambda$ is equivalent Early voltage associated with channel-length modulation effect; and K, m and n are device-specific coefficients.

Typical NMOSFET terminal I-V characteristics are shown in Figure 3.15. The voltage breakdown mechanisms, critical to ESD protection design, are multi-fold in nature. In long-channel device, the drain-to-source breakdown, BV_{DSS}, is caused by drain-to-substrate junction breakdown, BV_{DB}, governed by equations (3.6) and (3.7). In short-channel case, BV_{DSS} is caused by either parasitic lateral NPN breakdown or punchthrough. In the case of parasitic NPN of drain-substrate-source takes the role, BV_{DSS} follows

$$BV_{DSS} = BV_{CEO} \approx \frac{BV_{DB}}{\sqrt[n]{2}}\left(\frac{L_G}{L_{eff}}\right)^{\frac{2}{n}} \tag{3.22}$$

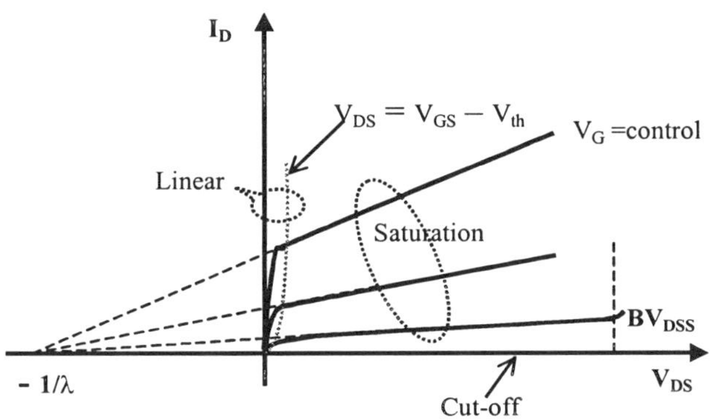

Figure 3.15 Typical common-source I-V characteristics for an enhancement mode NMOSFET transistor.

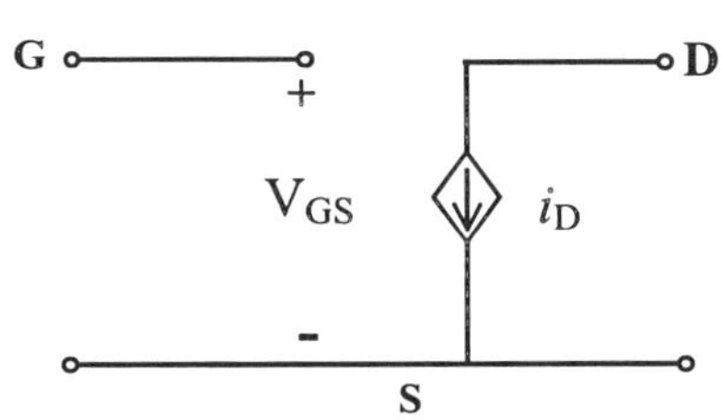

Figure 3.16 A large-signal model for NMOS transistor.

where BV_{CEO} is the open-base CE breakdown of the lateral NPN and L_G is drawn gate length, equivalent to base width of the parasitic NPN. Another breakdown mechanism that is particularly devastating to ESD protection is gate dielectric rupture, which deserves careful consideration in ESD protection design. A large-signal model for the NMOSFET is given in Figure 3.16.

3.4.2. ggMOSFET in ESD Protection Operation

One of the simplest MOS ESD protection device is the so-called grounded-gate NMOS (ggNMOS) structure, where the drain (D) goes to an I/O pad and the gate (G), source (S) and body (B) are shortened together to ground. Figure 3.17 illustrates its typical cross-section and equivalent circuit. The principle of a ggNMOSFET in ESD protection operation follows. As a positive ESD transient appears at an I/O pad, i.e., D, with respect to ground, the DB junction is reverse-biased all the way to its breakdown. Avalanche multiplication takes place and generates electron-hole pairs. Hole current flows into the ground via the B-region and build up a potential, V_R, across the lateral parasitic resistance R. Since the B and S regions are shortened together, V_R actually appears across the BS PN junction positively. As V_R increases, the BS junction turns on, eventually triggers the parasitic lateral

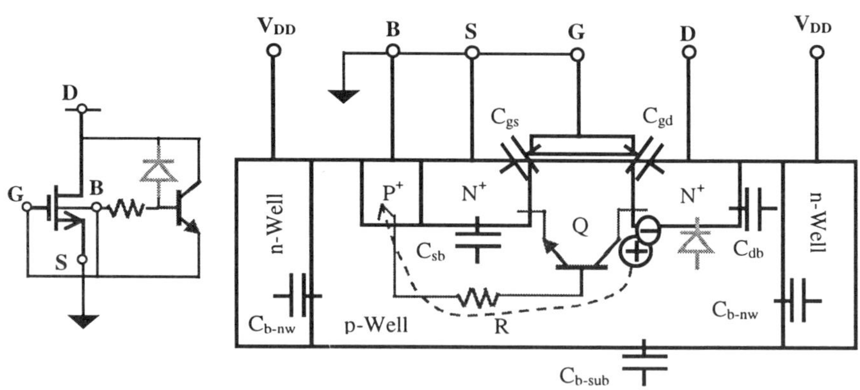

Figure 3.17 Typical cross-section and equivalent circuit for a ggNMOS ESD protection device.

NPN transistor Q (DBS). The rest of the story shall be the same as that told for BJT ESD protection device in Section 3.3. As a result, the ggNMOSFET is turned on at a triggering point, V_{t1}, under ESD pulse; moves into a snapback region to form a low-impedance discharge channel with low holding voltage, V_h; hence, discharges ESD transients safely. Clearly, the efficiency of this ggNMOSFET as an ESD protection structure is determined by the parasitic NPN BJT behaviour. Should a negative ESD pulse comes to the I/O pad, a forward-biased parasitic diode, BD, will take the role to shunt the transient. One of the main reasons for grounding the gate is to ensure "zero" leakage of the ESD protection structure under normal operations. This ggNMOS ESD protection structure has several advantages. Firstly, it provides an active discharging path, although in one direction only. Secondly, it is a natural option in CMOS technologies. There are disadvantages as well. For example, it cannot be included into circuit simulation due to its snapback I-V behaviour. Its application is also limited when being used as a complete ESD protection device because of the forward parasitic diode turn-on, which is too low for many high voltage applications. Figure 3.18 illustrates a typical ESD protection scheme for an I/O pad, where the I/O-to-V_{DD} protection resorts to a ggPMOS device. It is certainly not realistic to expect the same level of ESD protection from the ggPMOS because of the inefficiency of the parasitic PNP BJT. This is the reason that a ggPMOS transistor is normally much bigger than a ggNMOS

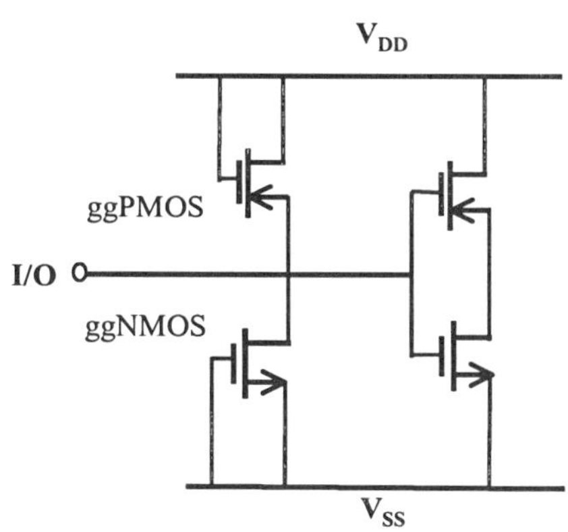

Figure 3.18 A complete ESD protection scheme for an I/I pad.

device in a design. Of course, there are different combinations available for complete protection of an I/O pad. Here comes the cheap saying again – *there are more than one route to Rome*. NMOSFET is one of the most studied ESD protection structures and has been the choice of ESD protection design for over one decade. A careful artistic work can achieve a fairly good ESD protection ratio of 16~20V/μm-width in 0.5μm technologies. However, to obtain high ESD protection, multiple-finger, or ladder, structures are needed. The detailed layout arts are pretty tricky. Modified MOS ESD protection structures will be discussed in Chapter 4.

A different flavour of a MOSEFT ESD protection utilizes a parasitic field-oxide MOS structure as shown in Figure 3.19. Assuming a piece of metal or poly-silicon runs over a thick field oxide (FOX) film between two N^+ regions, a parasitic MOSFET comes in effect at some biasing conditions. Then the same story holds: a parasitic lateral NPN in this parasitic MOSFET may be turned on to discharge ESD transients. Since the thick FOX film can survive much higher voltages compared to thin gate oxide and negligible off-state leakage is expected in a FOX MOS, one has an option to connect drain and gate together, resulting in lower triggering voltage and less leakage currents. One of the benefits is a reduced trigger voltage as compared to that in a grounded gate format due to raised gate bias, as will be discussed in Chapter 4 for a gate-coupled NMOS structure. This thick-gate MOSFET had been a choice of ESD protection for old technologies. However, it obviously not a charming option for area-efficient ESD protection because of, largely, its poor parasitic NPN operation.

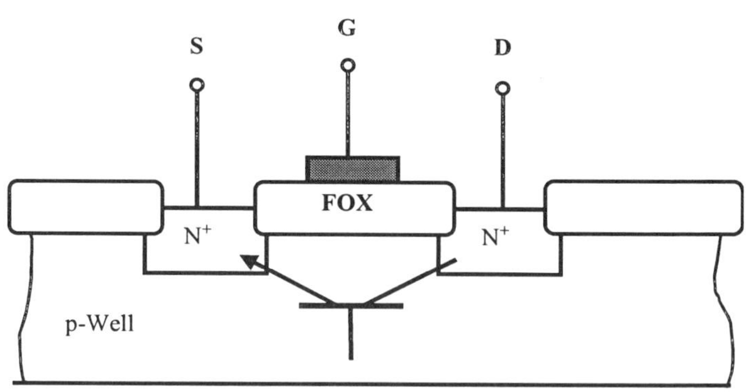

Figure 3.19 A cross-section for a thick-oxide NMOS ESD protection device.

3.4.3. MOSFET Parasitic Modelling

Parasitic effects associated with MOS ESD protection structures are fairly large in terms of absolute values. However, it is largely overlooked by IC designers. This design attitude must be corrected in today's IC design because such ESD-induced parasitic effects become relatively intolerable in RF and mixed-signal ICs and high-pin-count chips in VDSM (very-deep-sub-micron) regime. The parasitic models are discussed next, with detailed design consideration to be covered in Chapter 9.

Parasitic capacitances, C_{ESD}, can be analysed using the cross-section given in Figure 3.17. In ggNMOS ESD protection connection, gate-source overlap capacitance, C_{gs}, BS junction capacitance, C_{sb}, and body to guard-ring and substrate capacitances, $C_{b\text{-}nw}$ and $C_{b\text{-}sub}$, are negligible in small-signal *ac* model. This results in a parasitic capacitance model circuit dominated by gate-drain overlap and DB capacitances, C_{gd} and C_{db}. Consequently, $C_{ESD} = C_{gd} // C_{db}$ results, as shown in Figure 3.20.

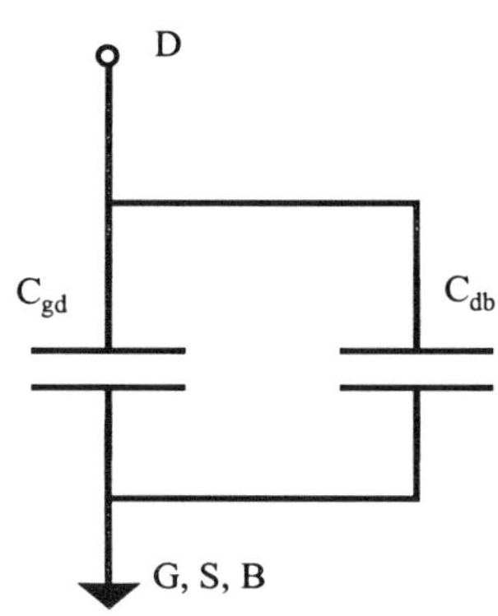

Figure 3.20 A parasitic C_{ESD} model for ggNMOS ESD protection structure.

Noise generators include the following: thermal noise due to channel conduction, $\overline{i_{nCh}}$; Flicker noise associated with i_D, $\overline{i_{nf}}$; induced-gate noise, $\overline{i_{ng}}$; noise generated by the distributed gate resistance, $\overline{i_{nr_g}}$; as well as thermal noises associated to series resistances, r_d and r_s, $\overline{i_{nr_d}}$ and $\overline{i_{nr_s}}$. These noises are given by the following formulas [4, 6, 7],

$$\overline{i_{nD}^2} = \overline{i_{nCh}^2} + \overline{i_{nf}^2} = 4kT\gamma g_m \Delta f + \frac{K i_D^{\alpha}}{f C_{ox} L_{eff}^2} \Delta f \tag{3.23}$$

$$\overline{i_{ng}^2} = \frac{16}{15} kT\delta\omega^2 C_{gs}^2 \Delta f \tag{3.24}$$

$$\overline{i_{nr_g}^2} = \frac{4kT}{3r_g} \Delta f \tag{3.25}$$

$$\overline{i_{nr_d}^2} = \frac{4kT\Delta f}{r_d} \tag{3.26}$$

and

$$\overline{i_{nr_s}^2} = \frac{4kT\Delta f}{r_s} \tag{3.27}$$

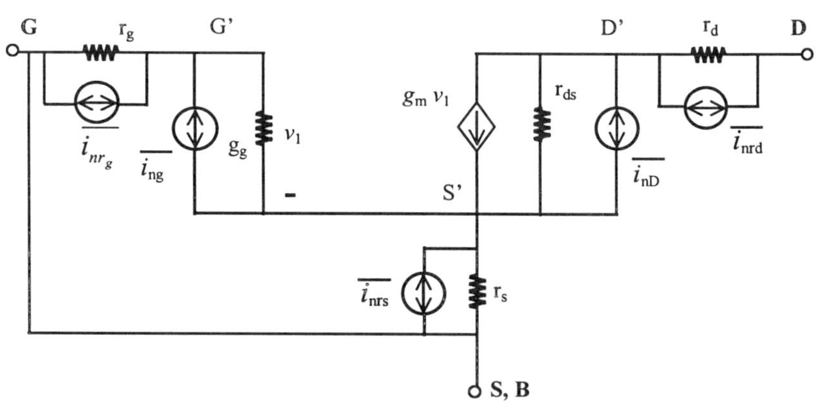

Figure 3.21 A noise model for a ggNMOS ESD protection device.

where γ, α, K, and δ are process and device specific coefficients, r_g is distributed gate resistance, and r_d and r_s are drain and source series resistances, respectively. As an ESD protection device that is normally off, the drain current, i_D, is a much smaller sub-threshold current. A complete noise model circuit is given in Figure 3.21.

3.5. SCR AS ESD PROTECTION ELEMENT

Silicon controlled rectifier (SCR) device may be one of the most efficient structure in terms of ESD protection because of its deep snapback I-V characteristics, which enables it to handle large current transients. However, it is also extremely tricky in designing functional SCR type ESD protection structures because of the potential latch-up phenomena. Understanding of SCR operation principles is certainly beneficial.

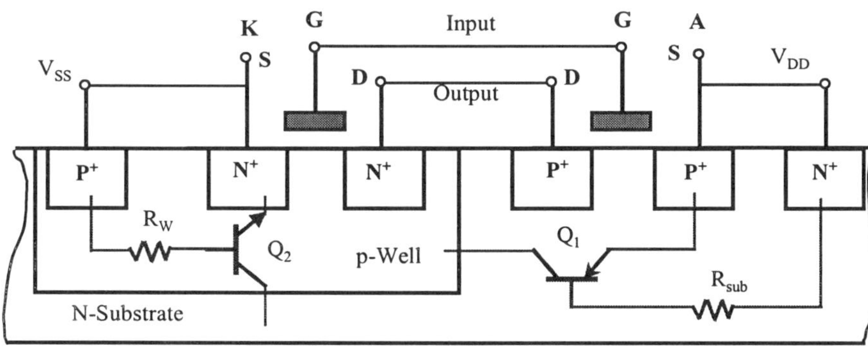

Figure 3.22 A cross-section for a CMOS inverter showing a parasitic SCR device inside.

3.5.1. SCR Device Physics

Although SCR structure might serve as an excellent ESD protection unit, it has been a nightmare to every IC professionals since CMOS technology was born. A parasitic SCR device may be found readily on CMOS IC chip, such as the one illustrated in Figure 3.22 for a classic CMOS inverter, where the parasitic SCR consists of a pair of parasitic lateral PNP, Q_1, and vertical NPN, Q_2, transistors, with their common-base/emitter current gains being α and β, respectively. Taking an intrinsic SCR structure shown in Figure 3.23 as an example. It is a four-layer, $P_1N_2P_3N_4$, device with an anode (A) and a

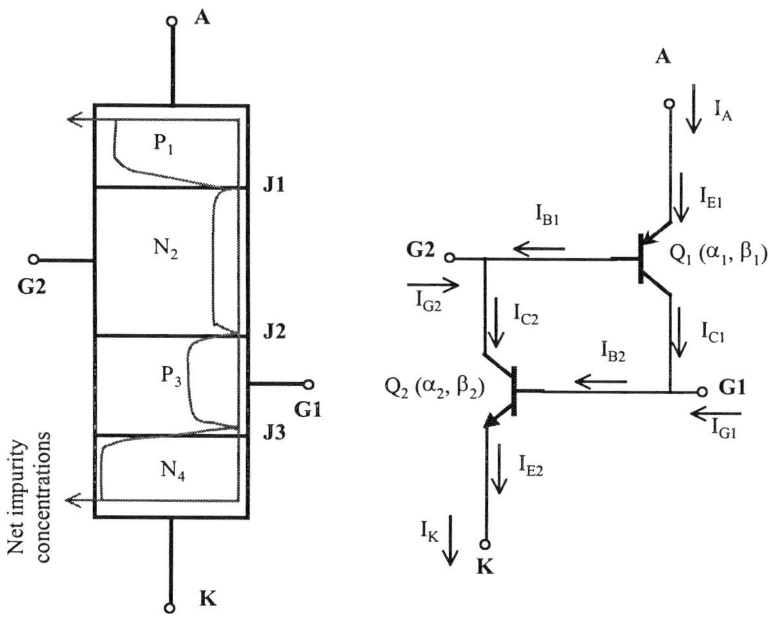

Figure 3.23 A generic SCR device and its equivalent circuit.

cathode (K) terminals and tow gates (G1 and G2). Typical net impurity concentration in a SCR follows a pattern of $P_1N_2P_3N_4 = P^+N^-PN^+$. A classic SCR terminal I-V characteristic is depicted in Figure 3.24, which comprises a reverse blocking region, ①; a reverse breakdown region, ②; a forward blocking region, ③; a negative resistance region, ④; and a forward conduction region, ⑤. Its operation principle follows. Starting from region ①, if a negative bias is applied from terminal A to K, junctions J1 (P_1N_2) and J3 (P_3N_4) are reverse-biased while J2 (N_2P_3) is positive-biased. There is no throughway, not even a toll way, existing for current conduction from terminal A to K, and SCR is in reverse blocking state. As negative V_{AK} increases, the reverse-biased junctions will be broken eventually, either via avalanche multiplication in depletion regions or punch-through breakdown, and driven into reverse breakdown region ②. In this illustration, and usually, the blunt of the V_{AK} will drop across J1 and cause avalanche in its depletion

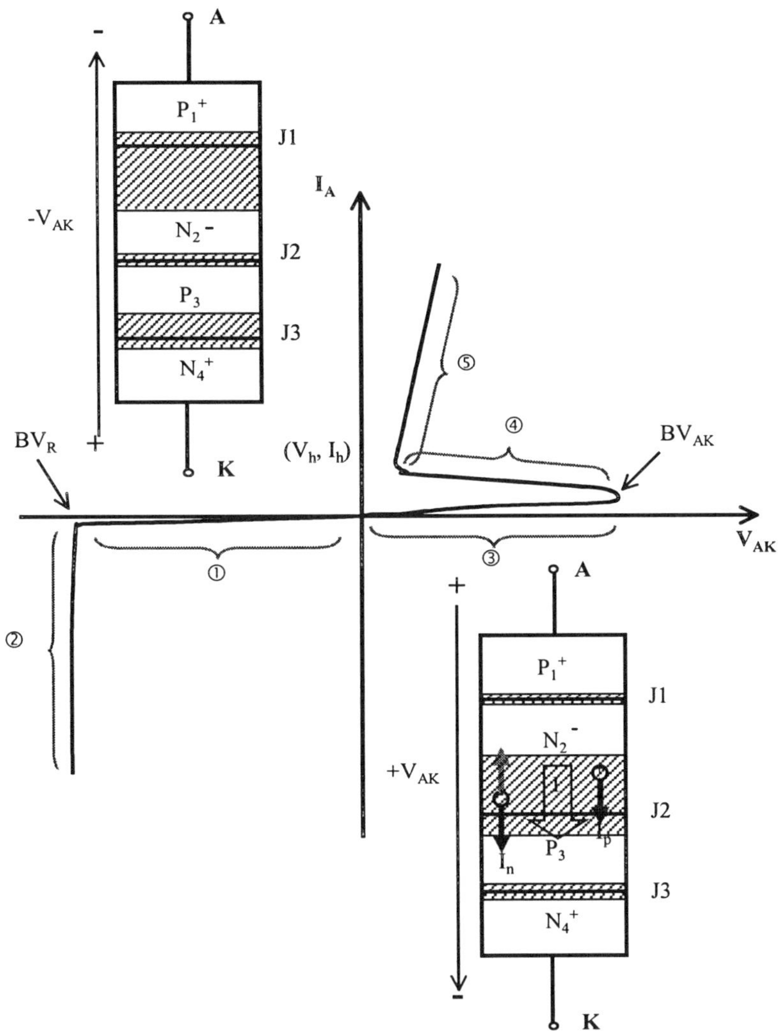

Figure 3.24 SCR I-V characteristics under positive and negative biases.

region. Considering multiplication in $P_1N_2P_3$, the reverse breakdown voltage, BV_R, is approximated by [3]

$$BV_R \approx BV_{DJ_1}(1-\alpha_1)^{\frac{1}{n}} \tag{3.28}$$

where BV_{DJ1} is reverse breakdown of diode J1 and n is a constant. In forward mode, when a small positive V_{AK} is applied at terminal A with respect to K, junctions J1 and J3 are forward-biased while J2 is reverse-biased. There is still no conduction channel between terminal A and K since J2 is off and it is referred to as forward blocking ③. Continuing increase in V_{AK} will result in breakdown in Q_1, which produces seeding current to trigger current regeneration in SCR. A brief regeneration mechanism is that, looking into the equivalent circuit of SCR in Figure 3.23, collector current of Q_1, I_{c1}, supplies base current, I_{B2}, for Q_2 and push it into active mode; in turn, collector current of Q_2, I_{C2}, sources base current, I_{B1}, for Q_1. Therefore, by intuition, as long as the current gain product greater unity, i.e., $\beta_1\beta_2 \geq 1$, (or, $\alpha_1\alpha_2 \geq 1$) [3, 9] the current regeneration sustains and a SCR device operates. The triggering of regeneration will move SCR through a negative resistance region, ④, into a low-impedance, high-current, low-voltage forward on state, ⑤. Now, look at what determines forward breakdown voltage, BV_{AK}. Although a mathematical genius might claim that it is right at the $dV_{AK}/dI_A=0$ point, a circuit designer would rather have some physical senses about it. First, one ought to know that there are different ways to trigger a SCR, including voltage-triggering by slowly stepping up V_{AK} or using a dV/dt transient, and current-triggering by injecting seeding currents from the gates (i.e., G1 and G2). Taking V_{AK}-induced avalanche breakdown as an example, where it is assumed $I_{G1} = I_{G2} = 0$. Using Ebers-Moll and Kirchhoff current law, one has

$$I_A = I_{E1} = I_{E2} = I_K, \tag{3.29}$$

$$I_{B1} = (1-\alpha_1)I_A - I_{CO1}, \tag{3.30}$$

and

$$I_{C2} = \alpha_2 I_K + I_{CO2}. \tag{3.31}$$

Since

$$I_{B1} = I_{C2}, \tag{3.32}$$

one gets

$$I_A = \frac{I_0}{1-\alpha_1-\alpha_2}, \tag{3.33}$$

where $I_0 = I_{CO1} + I_{CO2} << I_A$. One can get a good sense about the regeneration condition from this formula. Next to check what determines BV_{AK}. Assuming avalanche multiplication occurs in J2 depletion region as shown in Figure 3.24, where primary collector currents, I_{C1} and I_{C2}, come from the injected hole and electron currents I_p and I_n, both of which are multiplied by a multiplication factor, M (assuming $M \approx M_p \approx M_n$). The total current is

$$I = I_A = I_K = M_p I_p + M_n I_n = M(\alpha_1 I_A + I_{CO1}) + M(\alpha_2 I_K + I_{CO2}) \tag{3.34}$$

Hence, one has

$$\frac{1}{M(J2)} = \alpha_1 + \alpha_2 + \frac{I_0}{I_A}. \tag{3.35}$$

Alternatively, for avalanche breakdown in J2 depletion region, the multiplication factor cab be expressed by

$$M(J2) = \frac{1}{1-\left(\frac{V_{j_2}}{BV_{DJ_2}}\right)^n}, \tag{3.36}$$

where V_{J2} and BV_{DJ2} are reverse biasing and breakdown voltages for junction J2, respectively. Assuming the condition of $I >> I_0$, by equating M-factor equations (3.35) and (3.36), the forward triggering voltage can be approximated by

$$V_{t1} = BV_{AK} \approx BV_{DJ_2}(1-\alpha_1-\alpha_2)^{1/n}. \tag{3.37}$$

This formula indicates, quantitatively, that the SCR forward triggering is made easier due to parasitic BJT current gains, and is always less than, however not too much, that of reverse breakdown because both BV_{DJ1} and BV_{DJ2} are determined by lightly doped N_2 region. This break-over point is critical in designing SCR for ESD protection. Another important factor is its holding voltage, V_h, defined as the current needed to sustain forward conduction (or, latch-up in CMOS circuit) after triggering. Since in the ON state, Q_2 is in saturation, offering a remote direct-connection from P_1 to N_4, the SCR can be viewed as a P_1-I-N_4 diode in forward conduction. Hence, a $V_h \propto W/\tau_{eff}$ relationship shall hold, where W is the I-region width and τ_{eff} is effective lifetime. Well, so much is enough for an IC designer. Let's look into how a SCR works as an ESD protection structure next.

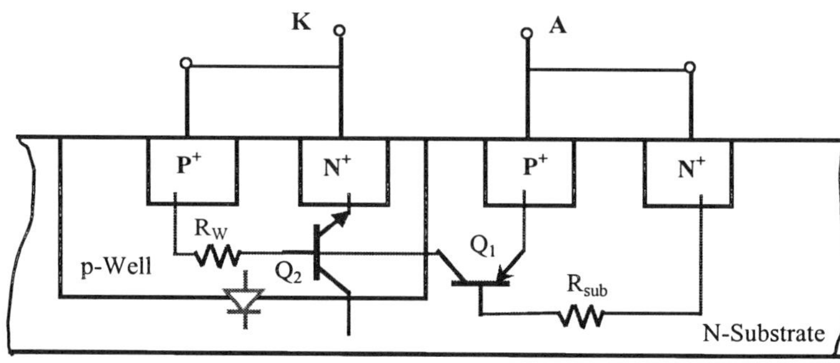

Figure 3.25 A simple SCR ESD protection scheme has the SCR structure connected into a two-terminal device.

3.5.2. SCR in ESD Protection Operation

A simple SCR-based ESD protection structure, as shown in Figure 3.25, is discussed here. More advanced versions of SCR ESD protection structure will be discussed in Chapter 4. One ESD protection scheme using SCR is to connect terminal A to I/O pad and terminal K to ground (GND, or V_{SS}) to protect against ESD pulses from I/O to GND. Based upon SCR operation, as a positive ESD transient appears at the I/O pad with respect to GND, the SCR is pushed into regeneration mode to form a low-impedance discharge channel to shunt the ESD current safely. Its triggering voltage is given by SCR forward breakdown, i.e., $V_{t1} \approx BV_{AK}$, which needs to be designed to a specific value. In the mean time, as SCR moves into ON state, I/O pad voltage will be clamped to a safely low level, i.e., the holding voltage, V_h, level. On the other hand, if a negative ESD pulse comes to I/O pad with respect to GND, a large parasitic diode of p-well/n-sub (J2) will be forward turned on and takes the charge. Hence, a SCR provides active discharge path in one direction and offers an asymmetric ESD protection solution. The main advantage of using SCR is its power to handle extremely large ESD pulses with a small size structure – a highly desirable feature for many IC chips that are parasitic and area sensitive. The disadvantages include latch-up possibility, incompatible with SPICE circuit simulation due to its snapback nature, and reverse diode operation that is not suitable for high voltage applications.

Similarly, based upon the generic SCR as shown in Figure 3.23, several factors must be considered in using a SCR ESD protection structure as shown in Figure 3.25, including parasitic resistance impacts, dV/dt effect and dI/dt limitation under ESD stresses. To start with parasitic effect by looking into Figure 3.25, since P_1, N_2 and P_3, N_4 are shortened together to anode A and cathode K, respectively, parasitic lateral resistances of N-substrate, R_W, and P-well, R_{sub}, must be included in analysis. Basically, as illustrated in Figure 3.26, shunting currents flowing through R_W and R_{sub}, I_W and I_{sub}, break up the closed regeneration loop of $I_{C2} = I_{B1}$ and $I_{C1} = I_{B2}$, resulting in degraded current regeneration. Consequently, it becomes harder to trigger the SCR into the ON state. Of course, if such kind of base-emitter

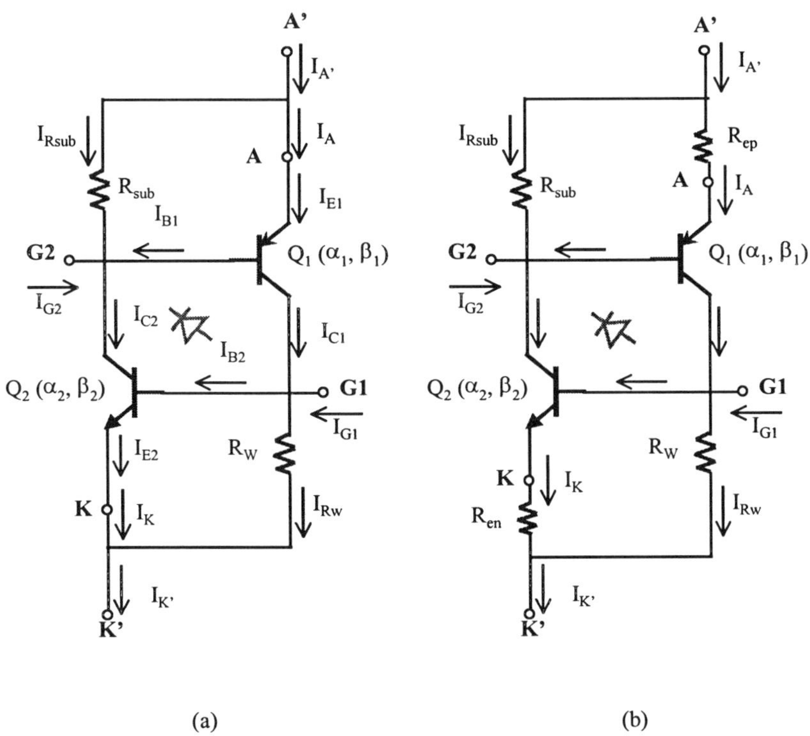

Figure 3.26 Schematics for SCR ESD protection structures including parasitic resistances.

short-circuit is inevitable, increasing R_W and R_{sub} shall reduce the B-E shunting currents and make SCR triggering easier. Quantitatively, the inner network between A and K in Figure 3.26a is the same as that in Figure 3.23. With the R_W and R_{sub} being taken into account, the avalanche multiplication analysis follows

$$I = I_{A'} = I_A + I_{sub} == I_{K'} = I_K + I_{R_W} \tag{3.38}$$

and

$$I = M_p I_p + M_n I_n . \tag{3.39}$$

Similar to the previous analysis and assuming I>> I_0, the triggering voltage can be approximated as

$$V_{t1} = BV_{A'K'} \approx BV_{DJ_2}(1 - \alpha_1 \frac{I_A}{I} - \alpha_2 \frac{I_K}{I})^{1/n} . \tag{3.40}$$

This result indicates that, due to the regeneration degradation effect, the triggering voltage V_{t1} (or, $BV_{A'K'}$) is greater than that of the generic SCR, BV_{AK}, without the parasitic resistances, R_W and R_{sub}. Further, the series resistances in the emitters of Q_1 and Q_2, r_{ep} and r_{en}, as shown in Figure 3.26b, have influences as well, as approximated by

$$V_{t1} = BV_{A'K'} \approx BV_{AK} + V_{r_{ep}} + V_{r_{en}} . \tag{3.41}$$

Because of the transient nature of ESD events, SCR operation under ESD may be different from classic SCR behaviours. Firstly, the dV/dt effect has strong impact on SCR triggering under ESD stresses. Using the similar avalanche analysis method, the displacement current of d(CV)/dt, flowing through junction capacitances, C, should be included into the total current as

$$I = M_p I_p + M_n I_n + I_d \tag{3.42}$$

where $I_d \equiv d(CV)/dt$ is the displacement current. Hence, a reduced triggering voltage is obtained as

$$V_{t1} = BV_{AK} \approx BV_{DJ_2}(1 - \alpha_1 - \alpha_2 - \frac{I_d}{I_A})^{1/n}. \tag{3.43}$$

Depending upon the magnitude of the displacement current, a SCR ESD protection structures, actually other types of ESD protection structures as well, may behave very differently from the design specifications, should the dV/dt factor not be included. This phenomenon was reported in several recent literatures [10, 11]. Secondly, a dI/dt effect may limit performance of a SCR ESD protection structure. An actual SCR triggering procedure is by no means a uniform event across the whole SCR junction areas. In reality, with the aid of seeding currents, initial turn-on occurs locally in a small area (*Anybody knows how small is this small?*) of a SCR junction and then spreads over the whole junction, which is characterized by a spreading velocity. Consequently, power density will not be the same across a junction. Hence, hot spots may appear in SCR junctions due to highly localized temperature overshoot. Qualitatively, a relationship of $P \propto \Delta T \propto dI_A/dt$ dictates. Therefore, pre-mature thermal damages may occur under ESD events due to this dI/dt limitation. One solution to this problem is to reduce the ratio of base width of Q_2 to diffusion length (W/L_{diff}). Alternatively, but less likely, a designer might have to bribe a testing engineer for slowing down the ESD zaps during ESD zapping tests. Another important thing to keep in mind in designing SCR ESD protection structures is that layout and placement of SCR on a chip are critical. As mentioned previously, current triggering is one way to turn on SCR. One ought to watch out substrate currents from surrounding elements and use special means, such as, double guard-rings, for isolation in order to avoid any accident triggering of SCR [12].

3.5.3. SCR Parasitic Modelling

Although SCR ESD protection structure features smaller size compared to other types of ESD protection structures, it is still necessary to look into its parasitic effects. Parasitic capacitances in SCR can be analysed using a typical structure shown in Figure 3.27, where P-well guard-ring is included to avoid accident triggering. Considering ESD operation nature of the SCR, only those reverse-biased junction capacitances are believed to be significant. The total ESD-induced parasitic capacitance network is modelled by a parallel capacitor network as shown in Figure 3.27, where C_1, C_2, and C_3 are associated with central P-well/N-substrate, and guard-ring P-well/N-substrate junctions, respectively. Therefore, the total ESD-induced parasitic

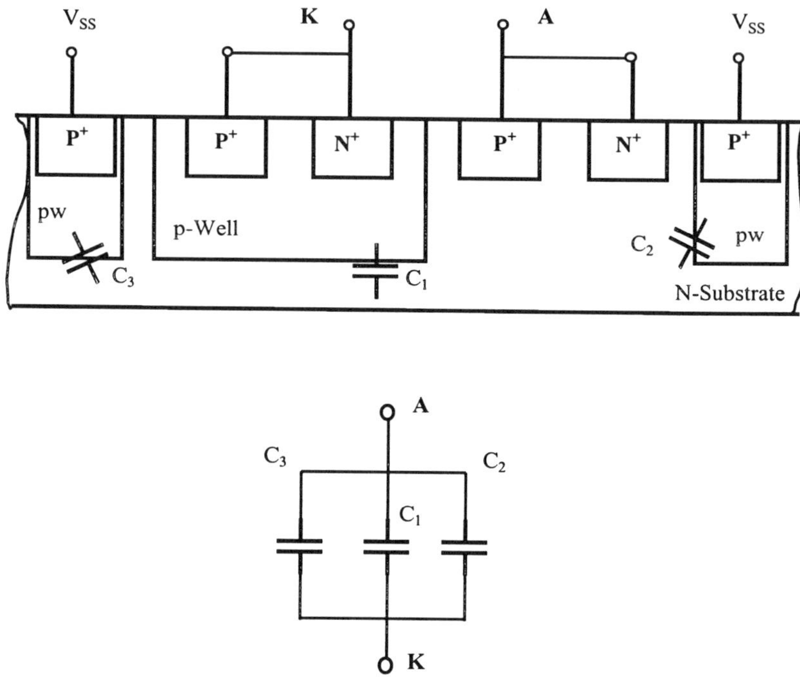

Figure 3.27 A typical parasitic capacitance model for SCR as an ESD protection device.

capacitance is given as $C_{ESD} \approx C_1//C_2//C_3$. ESD-induced noises can be modelled by the equivalent noise circuit shown in Figure 3.28 by treating SCR as a pair of BJT transistors, Q_1 and Q_2. The significant noise generators include thermal noises in all parasitic resistances, shot noises associated with emitter and collector junctions, as well as flicker noises in emitter junctions, as governed by the following equations for both Q_1 and Q_2 [4, 6, 7]

$$\overline{v_{nR}^2} = 4kTR\Delta f \tag{3.44}$$

$$\overline{i_{nC}^2} = 2qi_C\Delta f \tag{3.45}$$

and

$$\overline{i_{nb}^2} = 2qi_B\Delta f + K_1 \frac{i_B^a}{f}\Delta f \tag{3.46}$$

where R represents all the individual parasitic resistances (i.e., R_W, R_{sub}, r_e, r_b, and r_C for Q_1 and Q_2.), i_C and i_B are collector and base currents of Q_1 and Q_2, respectively. Of course, all currents are due to leakage. Since Q_1 and Q_2 share collector junction, a strong correlation factor is expected between i_{C1} and i_{C2} in this noise model.

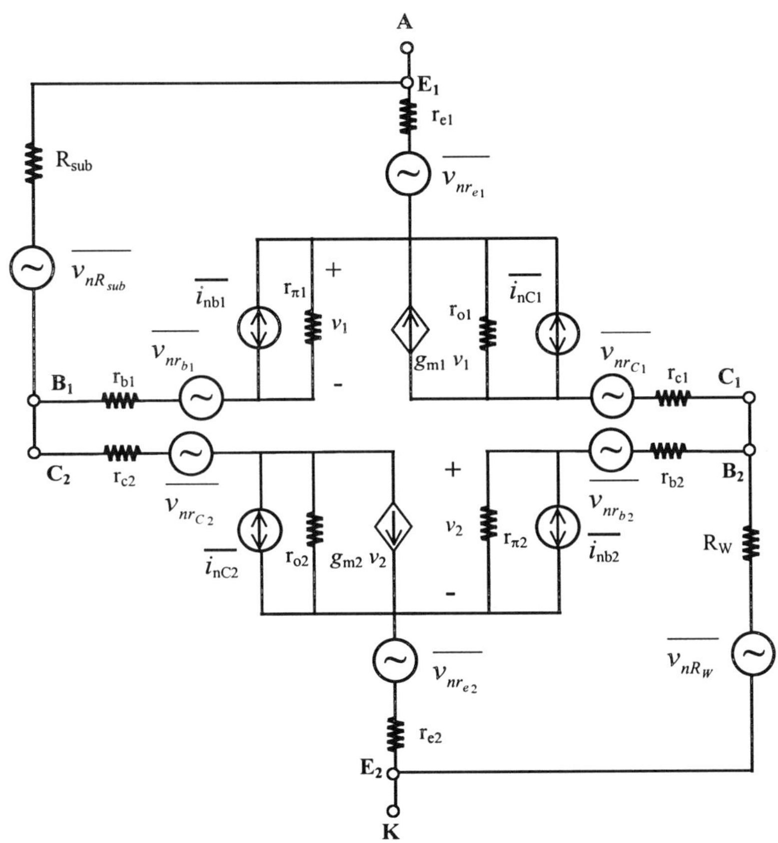

Figure 3.28 A ESD-induced noise model for a SCR ESD protection device.

3.6. SUMMARY

In summary, single-device ESD protection structures and their basic device physics are discussed in this chapter. The principles in ESD protection are to generate a low-impedance shunting channel to safely discharge large ESD currents and to clamp pad voltage to a sufficiently low level. Diodes are the simplest ESD protection devices, which can be included in circuit simulation. However, its low forward turn-on voltage makes it not suitable for many applications unless used in a diode chain format. Bipolar junction transistor is the basis for most active ESD protection structures including MOS and SCR types. BJT ESD protection structures in various versions are fairly effective ESD protection devices because of the large current gain. However, its snapback I-V behaviour excludes BJT from being simulated in simple SPICE way. NMOS is the most commonly used ESD protection structure in CMOS IC design. However, because of its low efficiency of the parasitic BJT inside, ggNMOS consumes large silicon sizes, hence, produces significant ESD-induced parasitic effects, which makes NMOS not a suitable solution for high ESD robustness, high-frequency, large pin count, and area-sensitive IC chips. SCR is one of the most efficient ESD protection structures, offering ESD protection of up to 80V/μm width [10], which makes it an attractive choice of future ESD protection for RF and mixed-signal ICs in VDSM regime that usually also demands extremely high ESD performance. However, success in designing SCR ESD protection largely depends on how well one can control the latch-up. Skilful layout and thoughtful placement of SCR structures on a chip play important roles in successful SCR ESD protection design. Equally important to achieving a super ESD performance level, it is imperative for circuit designers to include ESD-induced parasitic effects, both RC effects and noises, into chip design considerations, for which purposes, parasitic capacitance and noise models are introduced for typical ESD protection elements in this chapter. Circuit level ESD solutions based upon these single device elements will be discussed in Chapter 4.

REFERENCES

[1] Fong, Y. and Hu, C., "The Effects of High Electric Field Transients on Thin Gate Oxide MOSFETs", *Proc. EOS/ESD Symp*. 1987, pp. 252-257.

[2] Shockley, T., "The Theory of p-n Junction in Semiconductors and p-n Junction Transistors", *J. Bell Syst. Tech. 28,* 1949, pp. 435; *Electrons and Holes in Semiconductors*, D. Van Nostrand, Princeton, N. J., 1950.

[3] Sze, S. M., *Physics of Semiconductor Devices, 2nd Ed,* New York, Wiley, 1981.

[4] Gray, P., Hurst, P., Lewis, S. and Meyer, R., *Analysis and Design of Analog Integrate Circuits, 4th Ed*, Wiley, 2001.

[5] Miller, S. L., "Ionization Rates for Holes and Electrons in Silicon", *Phys. Rev., 105,* 1957, pp. 1246-1249.

[6] Lee, T., *The Design of CMOS Radio-Frequency Integrated Circuits*, Cambridge University Press, 1998.

[7] Massobrio, G. and Antognetti, P., *Semiconductor Device Modelling with SPICE, Second Edition*, McGraw-Hill, 1993

[8] Sze, S. M. and Chang, C. Y., *ULSI Devices*, New York, Wiley, 2000.

[9] Estreich, D. B., "The Physics and Modelling of Latch-up and CMOS Integrated Circuits", *PhD Dissertation*, Stanford University, 1980.

[10] Wang, A. Z., and Chen, C. H., "On a Dual-Polarity on-Chip Electrostatic Discharge Protection Structure", *IEEE Trans. Elec. Dev. Vol. 48, No. 5*, May 2001, pp. 978-984.

[11] Feng, H. G., "A Mixed-Mode Simulation-Design Methodology for on-Chip ESD Protection Design", *MS Thesis*, May, 2001.

[12] Wang, A. Z., Tsay, C. H., Bielawski, J. and DeClue, L., "Design Optimisation of a Practical ESD Protection Circuit by CAD: a Case Study", *Proc. IEEE Univ. Gov. Indust. Microelec.*, 1999, pp.116-119.

Chapter 4

ESD PROTECTION CIRCUIT SOLUTIONS

In Chapter 3, basic ESD protection mechanisms and device physics were discussed for single-element ESD protection structures, which are the foundation of all on-chip ESD protection design. These single-device structures have been dominating ESD protection solutions until early 1990s. However, modern ESD protection design is no longer a "device design" task. In addition, the old style of "one ESD protection structure for a chip" ESD protection design strategy does not hold any more in today's IC design. Advances in IC technologies make more robust but complex ESD protection circuits become attractive and feasible. Because of the facts that demands for ESD protection robustness increase continuously, complexity of IC functions requires customized ESD protection even within the same chip, and interactions between ESD protection structures and the circuit being protected become more and more important, development of advanced ESD protection solutions becomes a complex and challenging task that demands chip level consideration using a system approach. In this chapter, circuit level ESD protection solutions are discussed in great details.

4.1. INPUT ESD PROTECTION SCHEMES

Although many ESD protection structures can be used universally for input, output, power supply, and any other pads, design customisation is often needed for ESD protection solutions used at different pads. For example, a power clamp usually needs to deal with an ESD pulse in positive direction only, output pad ESD protection can be achieved using a large output buffer transistor itself, and input ESD protection may require low voltage clamping to protect CMOS gates. Nevertheless, the ESD protection principles remain the same, which are to create a low-impedance discharging path between every two pads to safely shunt ESD current transients and to clamp the pad voltages to a sufficiently low level. Typical ESD protection solutions for input pad are discussed in this section, of which, many may be used for other pads as well.

4.1.1 A Primary-Secondary ESD Protection Network

A classic primary-secondary ESD protection scheme is illustrated in Figure 4.1, which consists of a primary ESD protection structure, ESD_p, a secondary ESD protection unit, ESD_s, and an isolation resistor, R. A primary

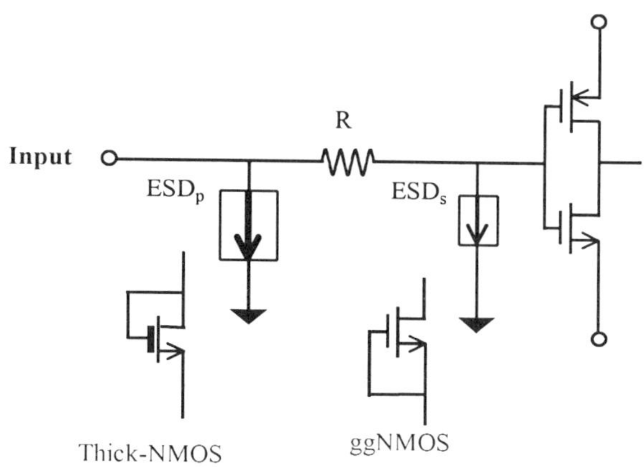

Figure 4.1 A typical primary-secondary ESD protection diagram consists of a primary protection unit, a secondary protection unit, and a current-limiting resistor.

ESD protection unit is designed to take the majority current of ESD transients. The role of a secondary ESD protection structure is to assist the turn-on of the primary that usually has relatively higher triggering threshold as well as to ensure sufficient low voltage clamping at a pad in order to protect a CMOS gate oxide. The main function of an isolation resistor is to limit the ESD current that flows into the core circuit being protected. In operation, as an ESD appears at input pad, the secondary ESD protection unit is turned on initially at a relatively lower voltage and the ESD current flows into ground via this shunting path. As voltage builds up at the left end of isolation resistor, the primary ESD protection unit will be triggered eventually. Obviously, the channel impedance of the primary path must be lower than that in the secondary channel in order for the primary protection unit to discharge most of the ESD current, which is the first key factor in designing such a protection network. Typically [1-3], a primary ESD protection unit can be realized by using a thick oxide NMOS, a SCR, or diode strings, where the capability of handling large ESD current transients is the most important specification. Other desired features include low holding voltage and small size. A ggNMOS can be a fairly good secondary ESD protection structure because it usually has low trigger voltage. In addition, the holding voltage of a ggNMOS should be safely low as far as CMOS gate breakdown is concerned. An isolation resistor should be large enough to ensure current-limiting function, however, without affecting IC speed performance. Low-parasitic poly-Si resistors may be used for low ESD case, while a diffusion-resistor with its heads in n-wells has better thermal behaviour for high ESD protection. Better ESD performance can be achieved by optimising layout; for example, current uniformity can be realized by rounding the corners of diffusion layers to avoid localized heating [4]. Interestingly, if a robust ESD primary (or, secondary) protection structure with both low trigger voltage and low discharge resistance available, often true in today's technology, one single such unit shall replace the primary-secondary network. In many cases, if a parasitic internal shunt path exists or the holding voltage of the ESD protection device is higher than an internal breakdown voltage, an isolation resistor is still needed. Keeping in mind that a simpler design is always preferred as long as ESD performance can be retained. Simplicity shall be the first golden rule in any engineering design.

4.1.2 Multiple-Finger ESD Protection Structure

Looking at any ESD protection layout using MOS protection structures, one can find that a big chuck of finger structures is usually used. The reason for that is to ensure high ESD protection by maintaining uniform current distribution, fairly similar to designing a large output buffer transistor. Because of its relatively lower ESD protection rating of a MOS protection structure, a fairly large device may be needed to realize high ESD protection level. For example, a 400μm wide NMOS structure may be needed in order to achieve 8kV ESD protection in HBM mode in a robust 20V/μm-width NMOS ESD technology. Unfortunately, it is not feasible to retain a wonderful relationship for the ESD protection level and the device size in practical design because of operation defects, such as current crowding, etc [5]. It is normally suggested that the channel width of a single MOSFET transistor, or preferably called finger length by an ESD designer, should be between 40μm to 100μm. Using a multiple-finger, or ladder, structure as shown in Figure 4.2 is a straightforward solution to this problem in design practices [6, 7]. However, it is still difficult to achieve the NMOS protection structure size to the ESD protection level scalability as expected, if cares are not exercised in both design and layout. This skewed finger-number-to-ESD-protection relationship is one of the most trouble-some design problem in designing the multiple-finger type MOSFET ESD protection structures [6, 7]. The reason is that Murphy's Law always prefers to turn on one MOSFET

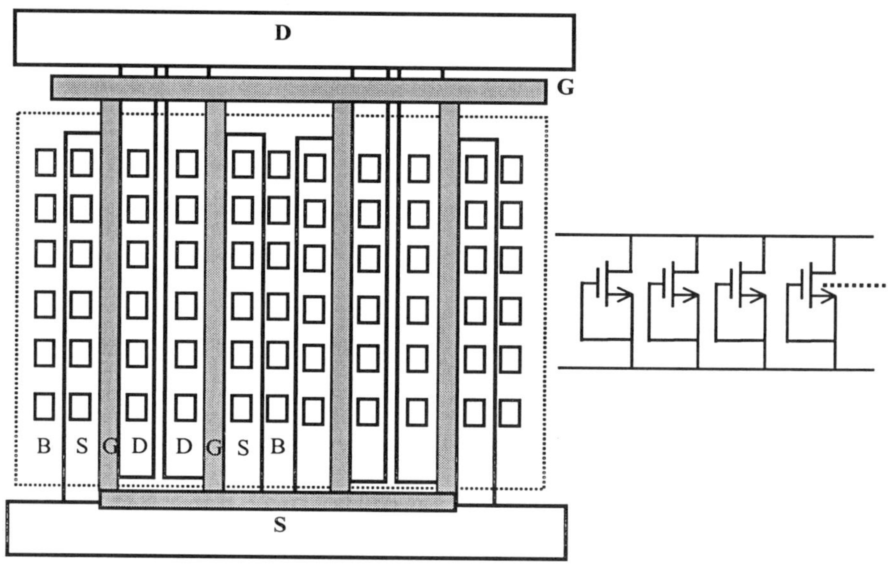

Figure 4.2 A sketch for a multiple-finger ggNMOS ESD protection structure.

finger ahead of any other ones. If the I-V characteristic of a ggNMOS ESD protection structure follows the solid-line as shown in Figure 4.3, where trigger voltage V_{t1} is greater that thermal breakdown voltage V_{t2}, i.e. $V_{t1} > V_{t2}$, it is easy to image the first turned on MOSFET finger starts to discharge the ESD transients and would reach to its damage threshold before any other fingers come into play. In such a case, simply increase of the number of MOSFET fingers would not offer a designer anything useful other than producing extra parasitic effects and consuming more silicon, which is particularly true in any silicided technology [8]. To resolve this problem, a

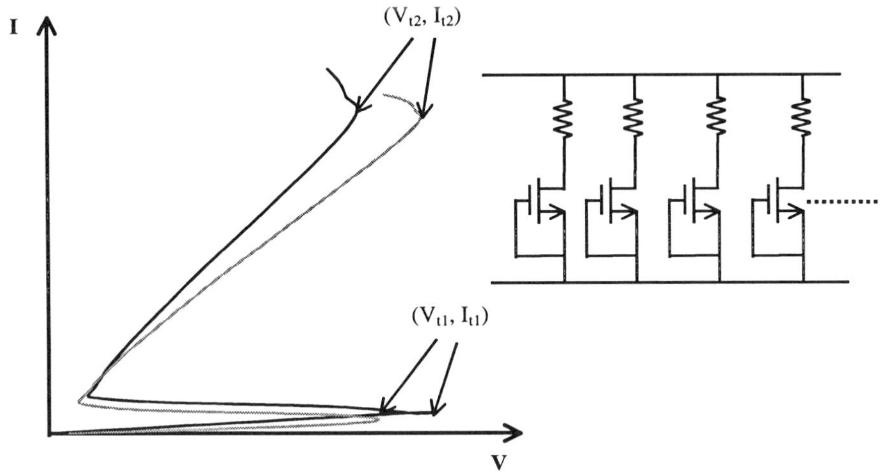

Figure 4.3 Different I-V characteristics result in different ESD performance for multiple-finger NMOS protection structures. A $V_{t1}<V_{t2}$ design (grey line) makes multiple-finger a useful structure while a $V_{t1}>V_{t2}$ design (dark line) breaks up the finger number to ESD protection scalability. Adding ballasting resistors improves current uniformity.

commonly accepted design principle is that one should make trigger voltage V_{t1} less than thermal breakdown voltage V_{t2}, i.e., $V_{t1} < V_{t2}$, as shown by the I-V curve in grey line in Figure 4.3. This way, before the *on*-finger has a chance reach to its thermal failure threshold, all other fingers can be turned on and join the game to discharge the ESD transients, at least in theory. Alternatively, one may insert ballasting resistor into each NMOS finger to ensure uniform turn on, as illustrated in Figure 4.3 as well. Since one usually has only limited cards to play in reducing V_{t1} in a specific technology, circuit techniques, such as the gate coupling technique, as to be discussed next, is often used in designing multiple-finger MOSFET ESD protection structures.

4.1.3 Gate-Coupled MOS ESD Protection Structure

Having known that an ideal design of $V_{t1}<V_{t2}$ is desired to make a multiple-finger NMOS ESD protection structure a useful solution, the next task is to implement this idea. A gate-coupling technique is one of the most

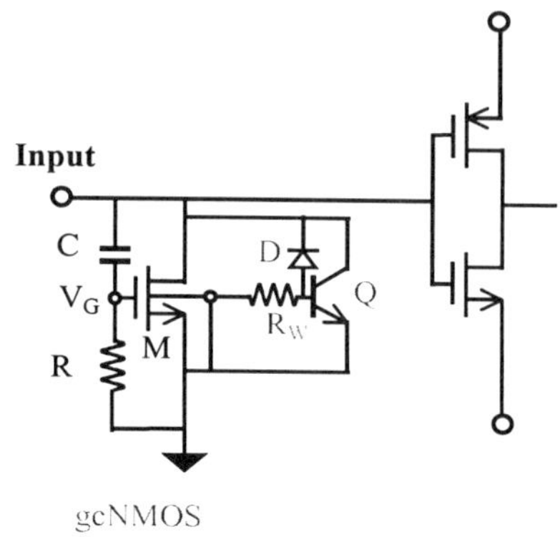

Figure 4.4 A gate-coupled NMOS ESD protection circuitry.

commonly used schemes in realizing the $V_{t1}<V_{t2}$ design. Figure 4.4 shows a typical gate-coupled NMOS ESD protection network, denoted by gcNMOS, where a RC coupling branch is connected to the gate of a NMOS protection device to bias up the gate. Its operation mechanism follows. As a positive ESD pulse appears at the input pad with respect to ground, the transient pulse is coupled onto the NMOS gate and readily raises gate voltage, V_G, to some level, which results in a reduced trigger voltage V_G. Assuming this theory holds, it would be possible to realize the desired $V_{t1} < V_{t2}$ condition by proper designing the gcNMOS network. Consequently, all fingers can be turned on before any one of them might have a chance to reach to its thermal damage point, V_{t2}, Even though the Murphy's Law might still pick up one sweetheart finger to turn on first. As long as a gcNMOS structure is turned on, the rest of the ESD protection story will be the same as that held for a ggNMOS ESD protection structure. The resistor, R, in series with the coupling capacitor, C, provides a discharge path for the gate after triggering in order to avoid long-fast leakage in NMOSFET. To understand the theory behind a gcNMOS protection operation, it is worth to repeat that an MOS ESD protection structure depends on its parasitic lateral NPN transistor in ESD operation as discussed in Chapter 3. That is, its DB junction breakdown occurs first, the generated hole current flows into the body terminal, gradually builds up a forward bias across the BS junction (i.e., V_{BE} of the NPN), and eventually turns on the NPN transistor. This story suggests that reduction of trigger voltage V_{t1} can be done by accelerating turn-on of the NPN. Obviously, one option will be to increase the substrate current, I_{sub}, which flows through the parasitic well resistor, R_W, resulting in a faster build-up of V_{BE} to turn on the parasitic NPN. Recall the classic MOSFET theory; several means may be utilized to increase the I_{sub} as a function of gate bias V_G. Firstly, there exists a relationship between substrate current, e.g., originated from avalanche breakdown at drain as shown in Figure 4.5, and gate bias, as illustrated in Figure 4.6a that remains true since the day of

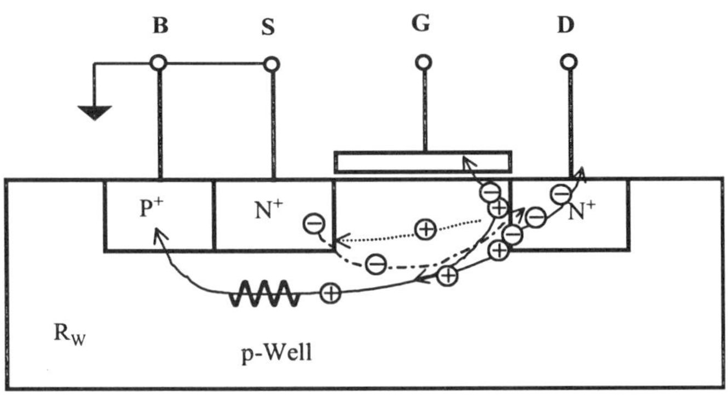

Figure 4.5 In a gcNMOS ESD protection structure, increase in substrate current from different sources, e.g., raising up the gate bias, results in decrease in trigger voltage.

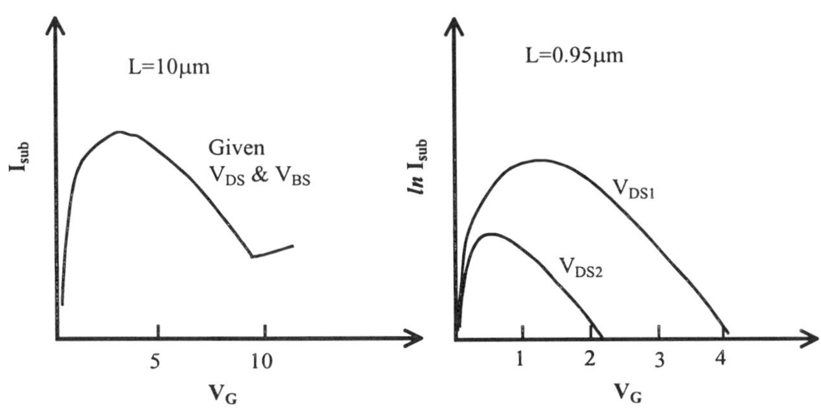

Figure 4.6 Typical bell shapes in I_{sub}~V_G characteristics suggest that proper increase in gate biases can accelerate triggering in MOS ESD protection structures: (Left) long-channel, (Right) short-channel.

Long-channel CMOS technology, which shows the I_{sub} peaking at some specific V_G value [9, 10]. Secondly, the famous hot-carriers come into play in short-channel MOSFETs. Well, assuming the threshold voltage shifting due to hot electron injection into gate oxide is not the concern for an NMOS ESD device, the hot holes will flow through the well, i.e., R_W, and be collected by the body terminal, hence contribute to the overall substrate current, I_{sub}. Magically, its $I_{sub} \sim V_G$ relationship also shows peaking in I_{sub} at certain V_G level, some suggests being $V_G \approx 0.4\sim0.5\ V_{DS}$, as illustrated in Figure 4.6b [9, 11-16]. In addition, hole current can reach the source region and be collected there directly in fairly short channel MOSFETs, causing forward-biasing the BS junction. Regardless which process dominates, as long as the BS junction is forward turned on, minority injection from source into body occurs, leading to the NPN turn on. Here comes the reason for an IC designer to know something about device physics: while substrate current should be suppressed normally for good reasons, such as, avoiding CMOS latch-up and decoupling substrate noises, it helps to trigger a MOSFET in ESD operation. This is how the trigger voltage V_{t1} of an NMOSFET can be reduced by the gate coupling technique. It is fairly obvious that proper design of the gcNMOS network is required to select an optimal gate bias V_G for a reduced V_{t1}. Design trade-offs exist. Gate oxide may be damaged if V_G is too high, even though below gate breakdown voltage BV_G, for too long a period. This is particularly true for very deep sub-micron IC design where the gate oxide is too thin to survive a voltage transient. Careful simulation work should be conducted in gcNMOS design, a topic to be covered in Chapter 8. Though a straightforward concept, design of a working gcNMOS protection network could be tricky in terms of selecting the values of coupling capacitor, C, and discharge resistor, R, the size of NMOSFET, RC timing, as well as doing the layout. Successes in designing gcNMOS ESD protection circuitry really depend on whether a designer can answer questions like "*why did you choose 10pF for the coupling-C, instead of 15pF?*" Based upon this concept, one can of course use different coupling techniques to reduce trigger voltage in design [17, 18]. It is worth note that the gate overlap capacitances in MOSFETs should be considered and a pad-oriented capacitor may be utilized in gcNMOS ESD protection circuit design.

4.1.4 BJT ESD Protection Network

Similar to junction diodes, BJTs, being in original, parasitic, lateral or vertical formats, are basic building blocks in most ESD protection structures.

While basic BJT ESD operation was discussed in Section 3.3, Chapter 3, countless BJT-based ESD protection sub-circuits were developed for various applications. This section introduces a few typical BJT ESD protection circuits.

As shown in Section 3.3, a resistor R is normally connected between base and emitter of an ESD BJT transistor to form an ESD protection network. Adversely, using the R results in higher trigger voltage V_{t1}. Such problem can be resolved by using a BJT ESD protection circuit shown in Figure 4.7. The principle follows. Recall that the base potential, V_B, building up via a resistor, R, is necessary to forward turn on the BE junction of a BJT in order to trigger its ESD protection operation. Conceptually, an external current source can be connected into the R-branch to increase V_B efficiently and rapidly. Consequently, this trigger-assisting current source accelerates ESD turn-on procedure compared to relying on avalanche breakdown currents. There are a variety of different ways to realize the trigger-assisting current sources. One of the easiest ones is using a Zener diode, usually available in BiCMOS technologies, as shown in Figure 4.7 [19]. As a positive ESD pulse appears at the pad, the Zener diode D_Z is reverse-biased to its turn-on threshold, V_{Dz}. The current flows through R, raises V_B, and turns on the BJT, Q. The trigger voltage, V_{t1}, approximately equals to V_{Dz}. Since majority of ESD transient will be absorbed by the ESD-optimised BJT transistor, a small Zener diode is adequate for trigger assistance. A coupling capacitor can also

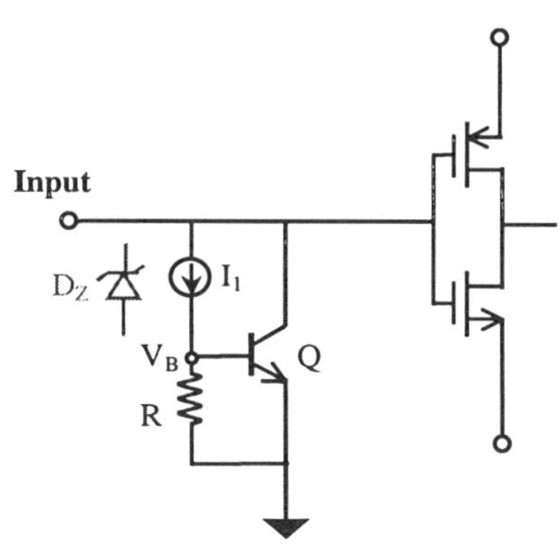

Figure 4.7 A typical BJT-based ESD protection uses an external current source to accelerate triggering. A Zener diode serves as a good trigger-assisting current source.

be used, replacing the Zener diode, to raise the base potential under ESD transients [20]. Figure 4.8a shows another BJT-based ESD protection circuit

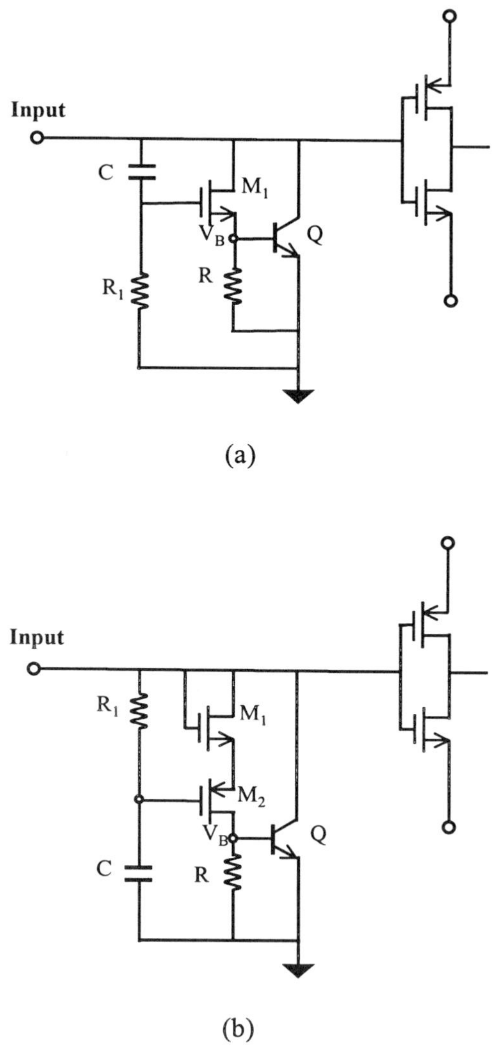

Figure 4.8 A BJT ESD protection circuit uses a gcNMOS as trigger-assisting current source: (a) gcNMOS trigger-assisting concept, (b) a gcNMOS–BJT protection circuit with a PMOS current breaker with sample values, C=10pF and R_1 = 2MΩ [21] (*Reproduced here by kind permission of the ESD Association and authors*).

topology, where a gcNMOS unit serves as the trigger-assisting current source. During an ESD event, the gate coupling effect turns on the NMOS, M_1, quickly, whose current, I_D, flows through R, builds up V_B, then turns on the ESD protection BJT, Q. In practices, the coupling capacitor, C, can be realized by the gate-drain overlap capacitance, C_{gd}, of M_1. However, excessive C_{gd} may also cause accidental non-ESD turn on of BJT Q due to noise coupling and significant signal variations if the RC time constant is not properly designed. One way to avoid such non-ESD turn-on is to insert a current breaker into the R-path. One such example ESD protection circuit is illustrated in Figure 4.8b [21], where PMOS, M_2, is connected between M_1 and R. During an ESD event, the gate of PMOS, M_2, is pulled down and M_2 is turned on immediately. The rest of the story will be the same as that for the circuit in Figure 4.8a. On the other hand, under normal operation, since M_2 is in off state, no current will flow through the R even in strong noise coupling situation. Of course, one of the most important tasks in designing such kinds of BJT ESD protection circuits is to properly select the values for those C's and R's. *But wait a moment, what do you means "proper"?!* Honestly, there is no universal answer to this question. The principle is that to make the RC time constants large enough compared to typical ESD pulse rise time, however, much shorter than that of the circuit being protected, so that the RC sub-net will only react to ESD pulses. This principle is a valid one because of the big difference in the timing between ESD pulses (less than 15ns) and ordinary circuits (often at ms level). Keeping this idea in mind, a circuit designer should know how to play around with these little R's and C's using SPICE simulation. However, the capacitance value required is often too big in such protection circuits, for example, a C=10pF was used in Figure 4.8b, which may disqualify it as an acceptable I/O protection solution in high-pin-count applications because it consumes too much silicon. Another example showing why an ESD protection unit can be called a "circuit" is given in Figure 4.9, where a fairly complex trigger-assisting network is used for a BJT ESD protection circuit [22]. Forget about all the confusing bells and whistles, the circuit is the same as that shown in Figure 4.7 except that a forward diode string serves as the external current source, instead of a Zener diode. This ESD protection circuit was designed for typical CMOS processes, where good quality Zener diodes are normally not available, or, officially supported, for low-voltage (3.3V) interface chips that have to communicate with their old higher voltage counterparts (e.g., 5V TTL logics). The BJT transistor, Q, is formed as a parasitic lateral NPN using the N^+ source/drains. In order for the 3.3V I/O pad to tolerate 5V interface signals without accidentally turning on ESD protection units, the trigger voltage, V_{t1}, must be set to above 5V with a safe margin, which is determined by the sum of voltage drops of the forward diode string and the

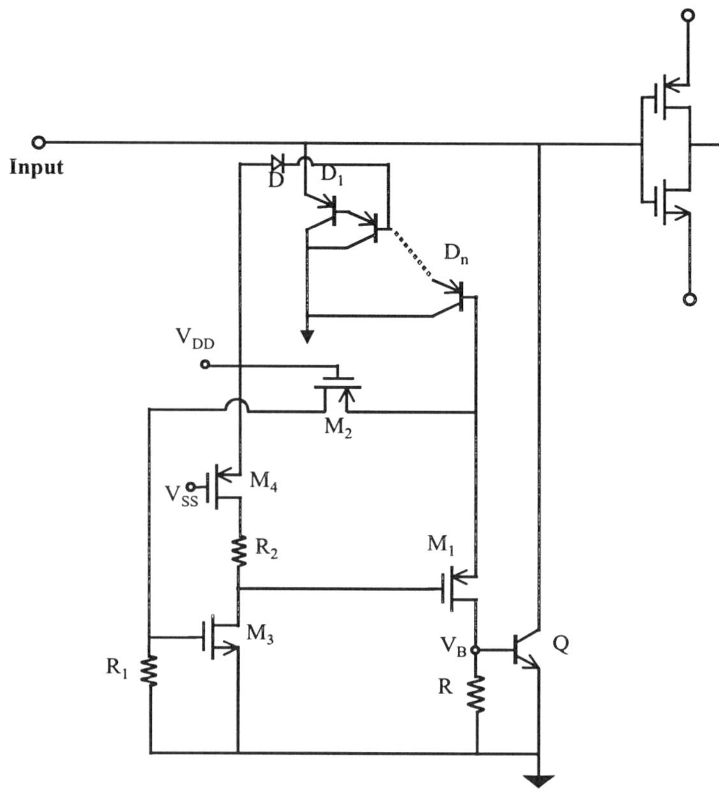

Figure 4.9 A BJT ESD protection circuit in CMOS technology uses a complex trigger-assisting sub-circuit [22] (*Reproduced here by kind permission of the ESD Association and authors*).

threshold voltage of PMOSFET, M_1. Following the previous discussions, during an positive ESD event, all the diodes and the PMOS M_1 will be turned on first, current flows through the R to trigger the BJT ESD structure Q. PMOS M_2 ensures full turn on of M_1 under an ESD pulse, where M_2 is turned on along with M_1 upon an ESD transient and its current, I_{D2}, flows R_1 to turn on NMOS M_3. Since PMOS M_4 is always on, turn-on of M_3 will pull V_{G1} to ground, hence locks the M_1 into the on state. While the logic sounds perfect in this story, a fair amount of efforts is expected in designing such kinds of complex ESD protection circuits. As an IC designer, one ought to be reminded of not losing yourself into the fun of design ESD protection circuits; instead, the revenues come from the IC chips only. When one argues about ESD protection design at circuit level, the real reason is that a single-device ESD protection structure may not work for a special circuit or circuit block. Otherwise, the golden design rule is always the simpler, the better.

4.1.5 SCR ESD Protection Network

One of the main hurdles impeding the use of SCR as ESD protection structure is that its trigger voltage is usually too high in normal CMOS processes for non-high-voltage applications. However, SCR emerges back as an attractive ESD protection option because of its high area efficiency that is highly desirable to parasitic-sensitive RF and high-pin-count ICs. Certainly, the high trigger voltage problem is getting worse in low-voltage very-deep-sub-micron (VDSM) applications. This is where ESD protection circuits, instead of stand-alone devices, come into play. Many SCR-based ESD protection circuit schemes, using trigger-assisting sub-networks, were devised to achieve low-V_{t1} SCR ESD protection. Starting with a non-circuit solution, Figure 4.10 shows a modified SCR cross-section in p-well/n-substrate technology, where an extra N^+ region (in grey) is placed across the inner p-well boundary. Recall the discussion given in Section 3.5, the trigger

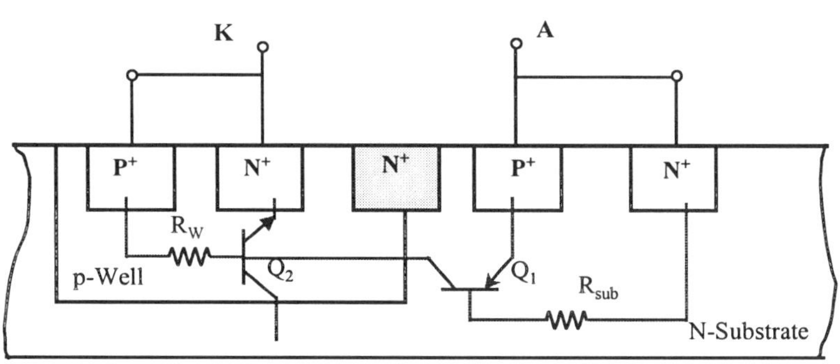

Figure 4.10 A cross-section for a lower-trigger voltage SCR ESD protection structure, where avalanche breakdown takes place at the N^+/p-well junction, instead of the n-substrate/p-well junction.

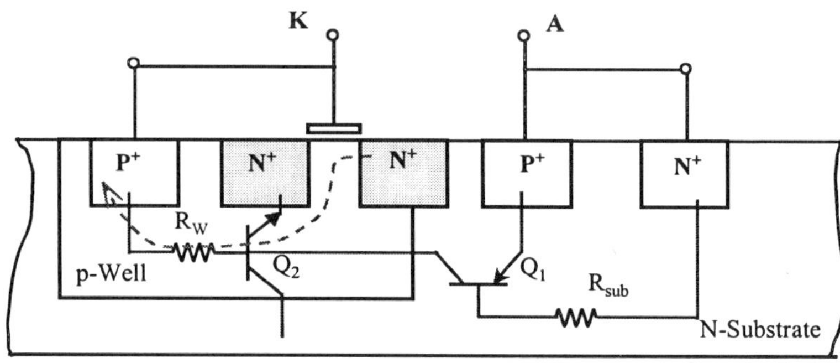

(a)

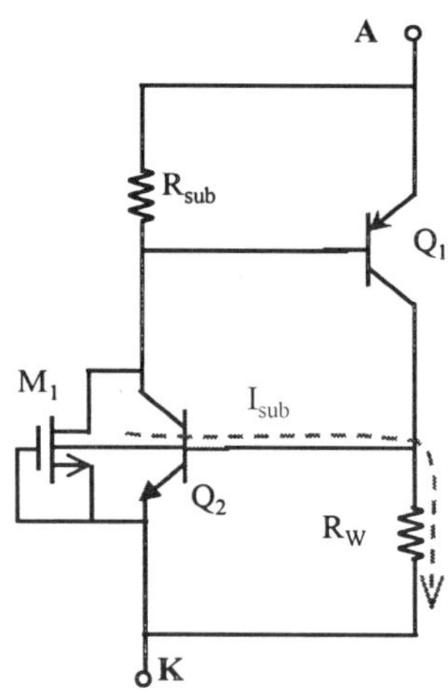

(b)

Figure 4.11 A ggNMOS-triggered SCR ESD protection structure reduces the triggering voltage: (a) a cross-section, (b) an equivalent circuit.

voltage of a basic SCR structure as shown in Figure 3.22 is determined by avalanche breakdown of the p-well/n-substrate junction whose value is usually very high. With the insertion of the N^+ region as shown in Figure 4.10, the avalanche breakdown now occurs the N^+/p-well at the inner p-well boundary, resulting in a much lower breakdown voltage. Hence, a relatively lower trigger voltage is realized. A simple circuit-mode low-voltage SCR protection structure is illustrated in Figure 4.11a, where an NMOSFET (in grey) is inserted into the cathode region. This small NMOSFET is actually connected as a ggNMOS as depicted by its equivalent circuit in Figure 4.11b. In ESD protection operation, an ESD pulse first turns on the ggNMOS due to drain junction breakdown and generates sufficient substrate current. The rest of the story is the same as that for a basic SCR structure, where the substrate current turns on the vertical NPN Q_2, hence, triggers the SCR structure. Assuming sufficient substrate coming from the ggNMOS, significant reduction in trigger voltage can be achieved. Implementation of this idea in an n-well/p-substrate process is given in [23]. One of the benefits for using this NMOS-triggered SCR circuit is that the trigger voltage can be tuned by varying channel length of the NMOEFET, at least in theory. A main problem associated with this protection circuit is that the thin gate oxide of ggNMOSFET is vulnerable to ESD transients because it sees an ESD pulse directly and might be damaged if the SCR does act swiftly [24]. Naturally, one may think of further reducing the trigger voltage by using a gate-coupling technique, i.e., including a gcNMOS trigger-assisting device, which has an even lower turn on threshold than that of a ggNMOS. Figure 4.12 shows an example schematic based upon this gate-coupling concept. This ESD protection circuit is actually in a complementary mode where a gcNMOS-triggered SCR and a gcPMOS-triggered SCR are used for I/O-to-V_{SS} and I/O-to-V_{DD} ESD protection, respectively. Making no mistake, the primary ESD discharging structure is still the SCR unit, where the gcMOS units are for easy triggering. Basically, the C_n-R_n-M_1 network (or, C_p-R_p-M_2) forms a gate-coupled MOS sub-circuit, which will be turned on first, and at a low voltage, upon an ESD stress. Then, the substrate current produced helps to trigger the SCR structure. A practical design based on this idea is given in [25]. It is certainly important to select proper values for the C-R-M network to ensure it operate only during ESD events, but not cause malfunction under fast signal processing due to possible C_n-R_n (or, C_p-R_p) short-circuiting. It is equally critical to make sure that the SCR's, not the MOSFET's, serve as the main ESD discharging devices. The former can be studied by a circuit designer by using SPICE, however, the latter concern may need mixed-mode electro-thermal simulation, to be discussed in Chapter 8, to evaluate the parallel competing ESD discharging channels if one does not want to rely on experience only. Once again, it is easy to make up a fairly nice looking logic

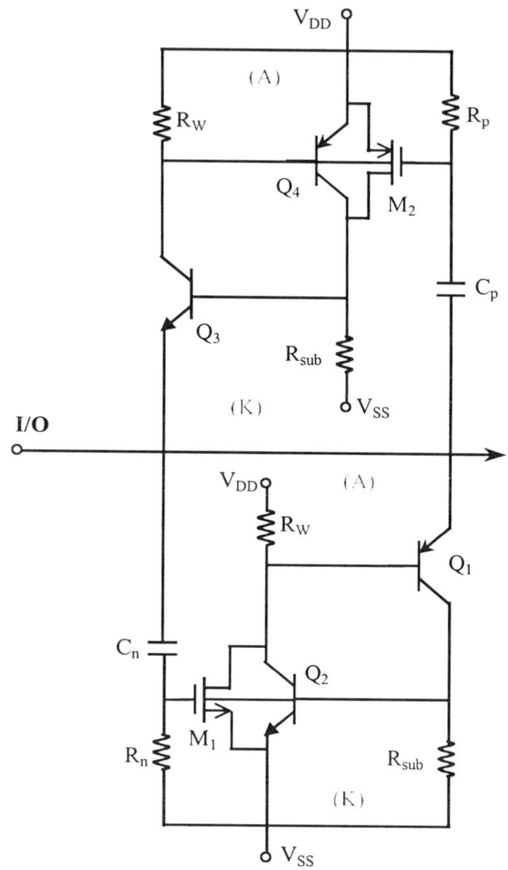

Figure 4.12 A complementary gcMOS-triggered SCR ESD protection circuit, in n-well/p-substrate process, uses gcNMOS/gcPMOS trigger-assisting units to achieve low trigger voltage [25] (*Reproduced here by kind permission of IEEE and authors*).

for a complex ESD protection circuit. However, to make it work properly is a different story. Another simple low-trigger SCR protection circuit is given in Figure 4.13 [26], where a Zener diode serves as an external current source. During a positive ESD event, the Zener diode, D_{Dz}, is turned on first, its current, I_{Dz}, flows through an external resistor, R_{ext}, and accelerates V_B building up for Q_2. Consequently, a lower trigger voltage can be realized

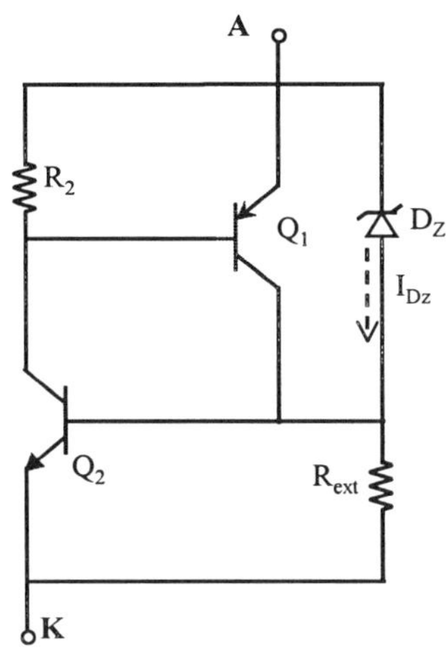

Figure 4.13 A low-trigger SCR ESD protection circuit uses a Zener diode trigger-assisting unit to reduce trigger voltage [26] (*Reproduced here by kind permission of IEEE and authors*).

for an SCR structure. Obviously, more smart SCR-based ESD protection circuits can be developed. This is where IC designers have a lot of edges. However, the golden rule is still that the simple the better, as long as it meets the specifications. A few more interesting compact protection designs will be discussed in Chapter 5.

4.2. OUTPUT ESD PROTECTION SCHEMES

Conceptually, ESD protection schemes for an output pad are similar to that for an input pin in a sense that both need to provide a low-impedance current discharging path with voltage clamping capability. Practically, large output buffer transistors themselves may have ESD protection capability as well. Therefore, output ESD protection structures can be classified as external ESD protection and self-protection. These two types of output pad ESD protection schemes are discussed in this section.

4.2.1 Dedicated Output ESD Protection Network

Like their input counterparts, output pads must be protected against ESD transients as well. Basically, all the ESD protection solutions discussed previously for an input pad can be used for effective output pad ESD protection. Nevertheless, due to their different functional roles and topologies, different design concerns should be taken care of in designing output ESD protection circuitry. Typical features associated an output pad include the following. An output pad is connected directly to source/drain diffusion regions in output buffers as oppose to insulating gates at input pads, hence, the large ESD transient currents flowing in/out internal circuit must be considered in design. Since an ESD protection structure is connected to S/D diffusion regions, one has to consider possible ESD failures due to junction breakdown and thermal melting, instead of concerning dielectric rupture in gate oxide at the input end. Because external currents can flow in/out output ports directly, special attentions must be paid to possibly noise-coupling induced accidental turn-on of ESD protection structures in noisy application environments. The noise injection/coupling effect becomes an issue because a great deal of ESD protection structures is based on parasitic bipolar transistor turn-on, which is usually enhanced by substrate current injection. Higher trigger current may help to suppress this problem [27]. In addition, if output buffer transistor self-protection scheme is used, some tweaks in design and layout of these buffer transistors must be made, which will be discussed in the next section.

As in input port ESD protection cases, the classic primary-secondary ESD protection scheme can be used for output pad ESD protection as illustrated in Figure 4.14. However, some special design considerations follow. For output ESD protection, a secondary ESD protection unit, ESD_S, may not be needed because gate oxide rupture is not a concern for output buffers, where pads are usually not connected directly to gates. In addition, the output buffer transistors may serve as the trigger device, hence, somewhat assuming the role of an ESD_S. Using an isolation resistor between the pad and buffer is usually not preferred because of the inevitable speed degradation due to direct current flowing in/out the buffer, although it may enhance ESD protection [28]. Therefore, design trade-offs are expected, or else, advanced ESD protection solutions are in demand. A complete output pad ESD protection scheme is given in Figure 4.15, where protection against all four ESD stressing modes, i.e., PS, NS, PD and ND, is ensured. Basically, all previously discussed ESD protection elements may be used in this output protection topology. For example, a ggPMOS and ggNMOS pair can be used as shown in Figure 4.15b. However, one must be sure exactly

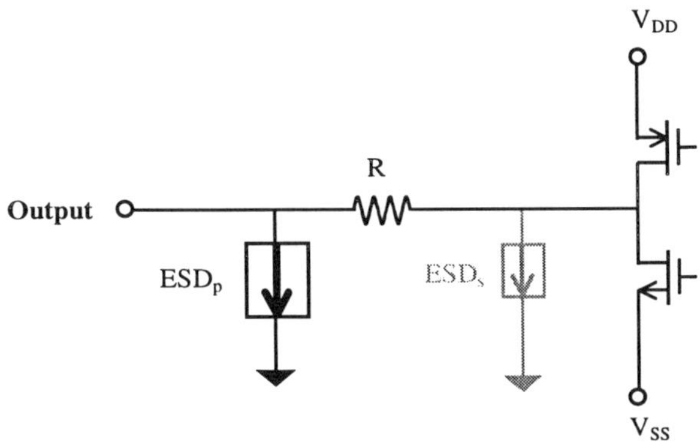

Figure 4.14 A primary-secondary output ESD protection scheme, where the pulldown NMOS may serve as the ESD_S and an isolation resistor, R, is optional.

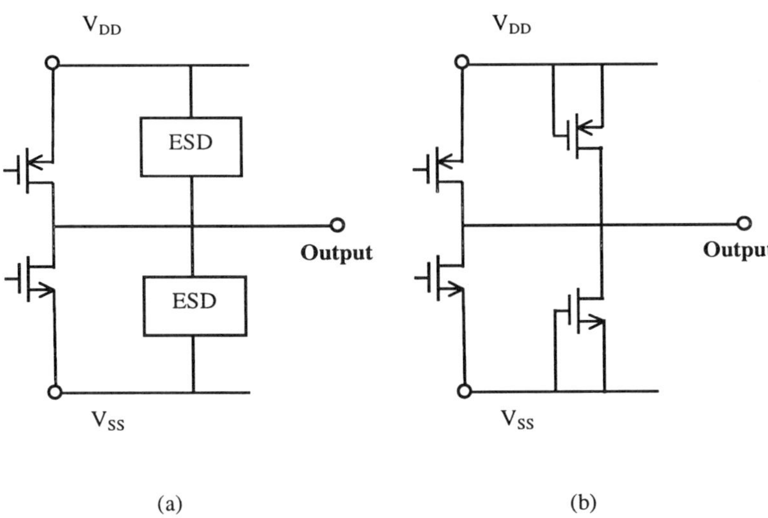

Figure 4.15 A complete output ESD protection scheme (a) and its implementation using a pair of ggNMOS/ggPMOS protection structures (b).

which MOSFET, the external ESD devices or the buffer transistors, will be turned on first under ESD pulses. This is not a trivial concern because both buffer MOSFETs and ggMOSFET ESD devices may use the same source/drain layers, and even channel lengths. A designer must properly design a lower trigger voltage for the ggMOSFET ESD devices. Otherwise, one has to ensure a much lower on-resistance for the ESD MOSFET in order to compete against the parallel buffer MOEFET. In fact, this parallel discharging competition is one of the most important design considerations in full-chip ESD protection design because pre-mature ESD failures are usually associated with a parallel weak link on a chip. Lose of discharging competition to a lower impedance parasitic shunting path in an output buffer block is often the root case of early ESD failure. Other ESD protection structures used for Figure 4.15 scheme include dual-diode [29] and SCR [30], etc. Another special consideration for output ESD protection design is that the pulldown NMOSFET is normally the most ESD-vulnerable device. It is believed that if this NMOSFET survives an ESD pulse, the whole buffer will survive it too [4]. Hence, many partial output ESD protection circuits were designed accordingly. Figure 4.16 shows an output protection circuitry,

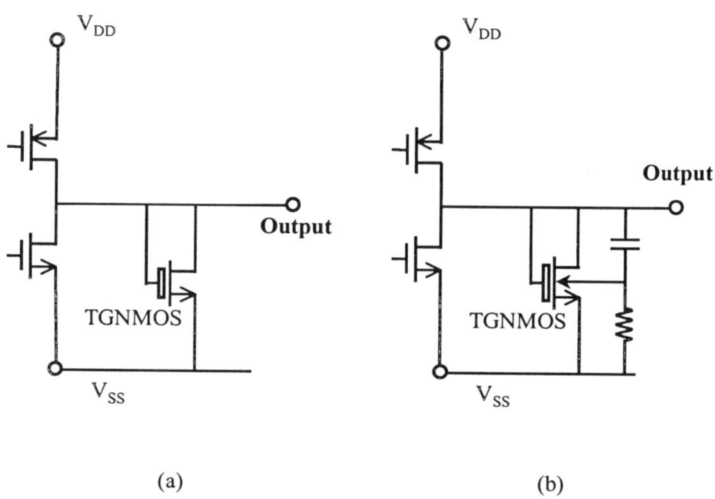

Figure 4.16 A TGNMOS ESD protection scheme protects the pulldown NMOEFET transistor in an output buffer.

where a thick-gate NMOS (TGNMOS) is placed between the pad and ground. Since a TGNMOS normally has a high trigger voltage, a well-coupled TGNMOS design, as shown in Figure 4.16b [31], was reported to reduce the trigger voltage. Similar to a gcNMOS, the ESD transient is coupled into the p-well of the TGNMOS via the RC branch that provides extra body current to turn on the lateral NPN within the TGNMOS. Another example using lateral NPN protection unit is shown in Figure 4.17, where a

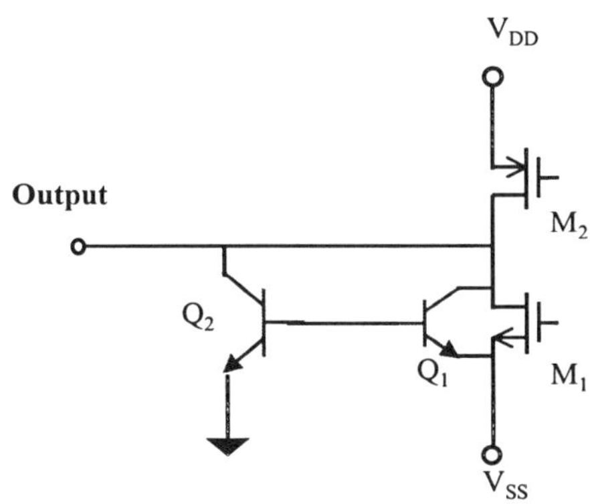

(a) A schematic.

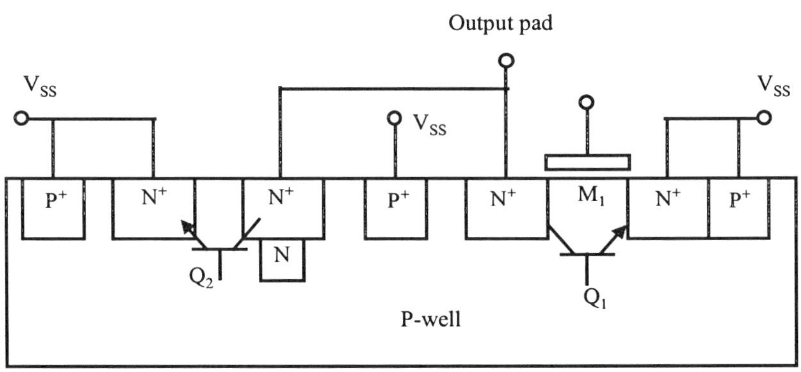

(b) A cross-section

Figure 4.17 A lateral NPN BJT output ESD protection scheme features a base width of Q_2 narrower than the channel length of the buffer NMOS Q_1.

lower trigger voltage for the internal NPN ESD protection device, Q_2, is ensured by making the channel length of the pulldown NMOS transistor a little longer than the lateral base width of the NPN, as shown in Figure 4.17b [32].

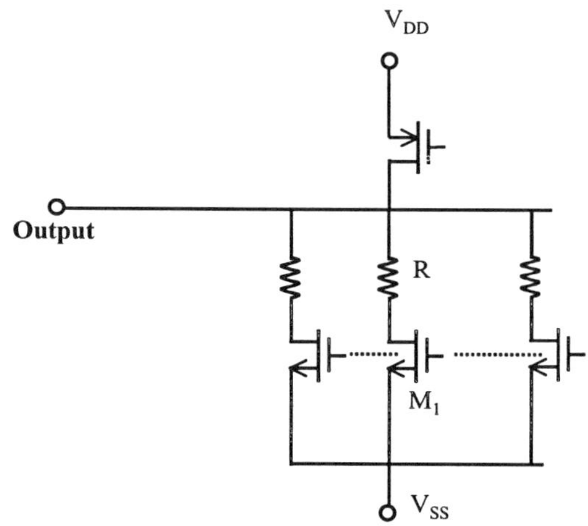

(a) A schematic.

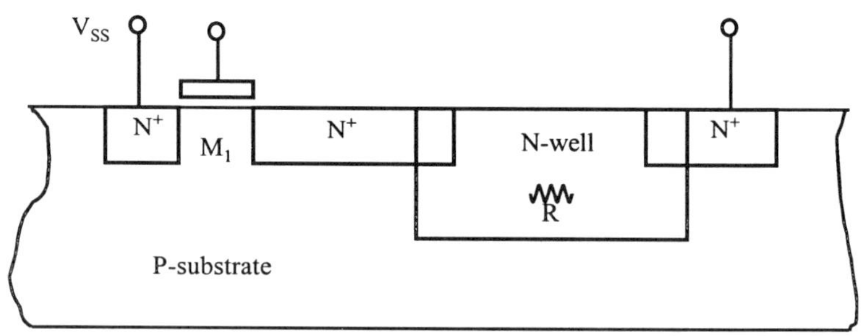

(b) A cross-section.

Figure 4.18 A self-protection output buffer NMOSFET uses an n-well extension resistor as ballasting resistor to ensure uniform triggering.

4.2.2 Self-Protection of Output Stages

As mentioned earlier, the buffer transistor itself can be used as a self-protection device because it generally has very high current handling capacity and is very large in size. Apparently, these large buffer transistors should be modified somehow to meet ESD operation requirements without losing original driving functionality as a buffer. This usually means design trade-offs for these output buffer transistors. Of course, there is no free lunch anyway. One major design concern for an output buffer MOSFET is that multiple finger layout is generally used and one must ensure uniform turn-on and current/heat distribution across all fingers, just like in a ggMOS ESD design case. Gate-coupling technique is useless in buffer MOSFET design because it will alter the buffer circuitry completely. One common option is to place ballasting resistors between output pads and drains of NMOSFETs. Figure 4.18 depicts one such design, where an n-well diffusion resistor is inserted between the internal drain region and external one [33]. The n-well diffusion resistor is preferred because it provides sufficient resistance with limited size and has good vertical heat dissipation capability. However, if silicide technique is used, this diffusion resistor may lose its value unless a field-oxide pinched diffusion resistor is available. Or else, pollysilicon resistor may be used instead, which, however, usually has poor thermal property. There is no doubt that layout plays an important role in designing such self-protection buffer transistors. This aspect will be discussed in Chapter 7.

4.3. POWER CLAMPS

A power-clamping device is needed for two reasons. First, a power bus may experiences ESD surges that may cause internal circuit failures. Second, an efficient power clamping device working with single-direction I/O ESD protection devices can provide full ESD protection between any two pads on a chip. The latter applications will be discussed in Chapter 5 for whole-chip ESD protection schemes. As a power clamp, an ESD protection structure is placed between V_{DD} and V_{SS} (or, Ground) pads, which forms a low-impedance current discharging path and clamps the power bus to a sufficient low holding voltage level during an ESD event. Power clamps are often placed at the four corners on a chip for efficient ESD discharging. However, the number of power clamps and their placement should be determined according to the layout and transistor density of an IC chip to ensure short, low-impedance and evenly allocated discharging paths. In case multiple power supplies are used on a chip, power clamps should be placed between local power buses as well. This scheme will be discussed in Chapter 5 for whole-chip protection solutions. Conceptually, any ESD protection devices can be used as power clamps. Typical power clamps are discussed next.

4.3.1 NMOS Power Clamp

NMOS structures are classic ESD protection solutions that also work well as power clamps. Figure 4-19a shows a typical ggNMOS power clamp. A lower trigger voltage for the power clamp is critical to ensure the power clamp turns on before any parasitic internal device does under an ESD pulse. Hence, the trigger voltage of a power clamp should be tuned sufficiently low, however, higher than the power supply voltage with an adequate safe margin to avoid accidental latching due to normal power supply fluctuation during normal circuit operation. A positive ESD surge at a power bus, the main concern for power line protection, will be discharged safely through the parasitic NPN in the ggNMOS. The parasitic p-well/N^+ diode will take a negative ESD pulse. Since the trigger voltage of a ggNMOS is low enough and no gate oxide is in direct connection with the clamp, a secondary ESD protection device is not necessary in this scheme. However, if a thick-gate (field oxide) NMOS structure is used as the power clamp, as shown in Figure 4-19b, a secondary ggNMOS ESD protection device and an isolation

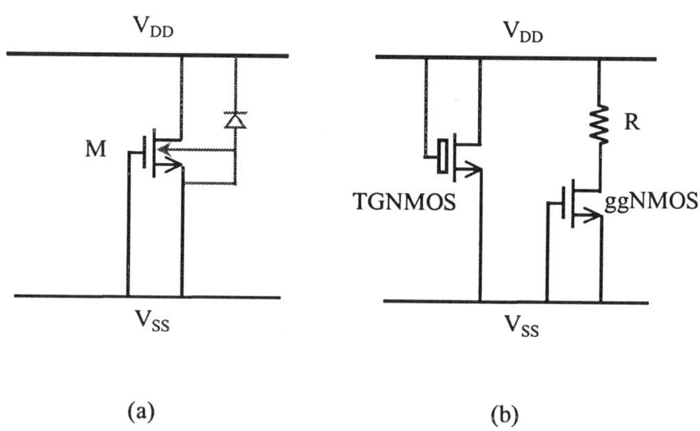

Figure 4-19 Power clamps using ggNMOS (a) and TGNMOS (b) structures.

resistor are required for efficient ESD protection. One main disadvantage in using NMOS power clamps is that the device size is usually very large. Though carefully designed finger structures may be used to ensure uniform ESD discharging, the big size will consume substantial silicon asset, produce significant parasitic effects and cause problems in chip layout.

4.3.2 SCR Power Clamp

In terms of power clamping, SCR type structures are very advantageous, should latch-up possibility be eliminated. A conventional SCR is usually not an option because of its high trigger voltage, which often results in turn-on of internal weak links, instead of the dedicated SCR clamping structure. Various modified low-voltage SCR structures can be used as power clamps instead. For example, Figure 4-20 shows an NMOS-triggered SCR clamp, which has a sufficiently low trigger voltage, as discussed in Section 4.1.5. No secondary protection device is necessary because of the low trigger voltage. The main advantages of an SCR clamp are its small size and high current-handling capability, which solve the problems associated with NMOS clamps as discussed previously.

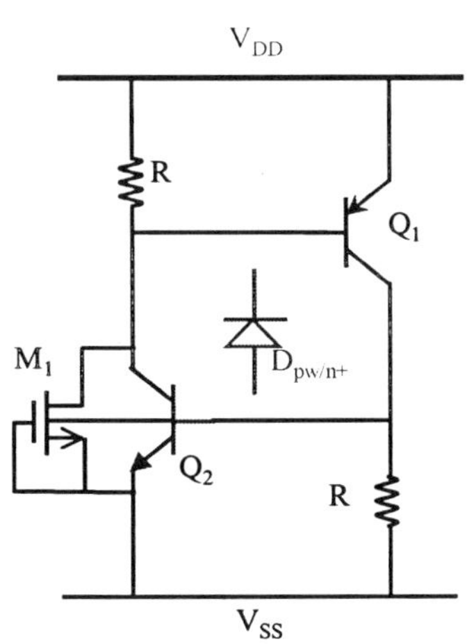

Figure 4-20 A power clamp uses an NMOS-triggered SCR Structure.

4.3.3 Diode String Power Clamp

As always, the golden rule in ESD protection design is the simple, the better, provided adequate ESD protection being assured. In this sense, forward diodes should be the choice. However, even low-voltage technologies (e.g., 1.8V) cannot use single forward diodes as power clamps, because its forward turn-on voltage is too low (i.e. ~0.65V for silicon diodes). One simple solution is to use a diode chain to realize a sufficient high turn-on threshold. This type of diode chain concept has been used in practical designs [34, 35]. However, issues arise in diode chain applications that demand careful design trade-offs among precise turn-on voltage control, leakage reduction and ESD performance. Assume a six-diode chain is used as a power clamp between the V_{DD} and V_{SS} pads in an n-well technology, where a diode is formed by a P^+/n-well junction. The diode chain (D_1 to D_6) is electrically a PNP transistor chain (Q_1 to Q_6) as depicted in Figure 4-21. It is worth to point out that the diode chain is used as forward-mode diode string for high current handling capacity. Let's first estimate the trigger voltage, V_{t1}, of the diode chain. Assuming identical diodes throughout and if ideal, clean diodes being used, the V_{t1} will be the sum of the forward turn-on voltage, V_D, of the six diodes, i.e.,

$$V_{t1} = \sum_{i=1}^{6} V_{Di} = 6V_D \quad . \tag{4.1}$$

This results in a V_{t1} of about 3.9V, a good value for a 3.3V V_{DD} clamp. Unfortunately, this nice linear add-up formula does not hold for an IC diode chain because of the Darlington amplification effect. From the Shockley equation for a PN junction diode, a forward turn-on voltage is given by,

$$V_D \approx nV_T \ln \frac{I_D}{I_S} \tag{4.2}$$

where V_T is thermal voltage, and I_D and I_S are diode current and saturation current, respectively. For a PNP-based BE junction diode as shown in Figure 4-21 and assuming identical Q's with a current gain of β, the following derivation holds. Starting from D_1 (Q_1),

$$V_{BE1} = V_{D1} \approx nV_T \ln \frac{I_{D1}}{I_S} = nV_T \ln \frac{I_{E1}}{I_S}, \tag{4.3}$$

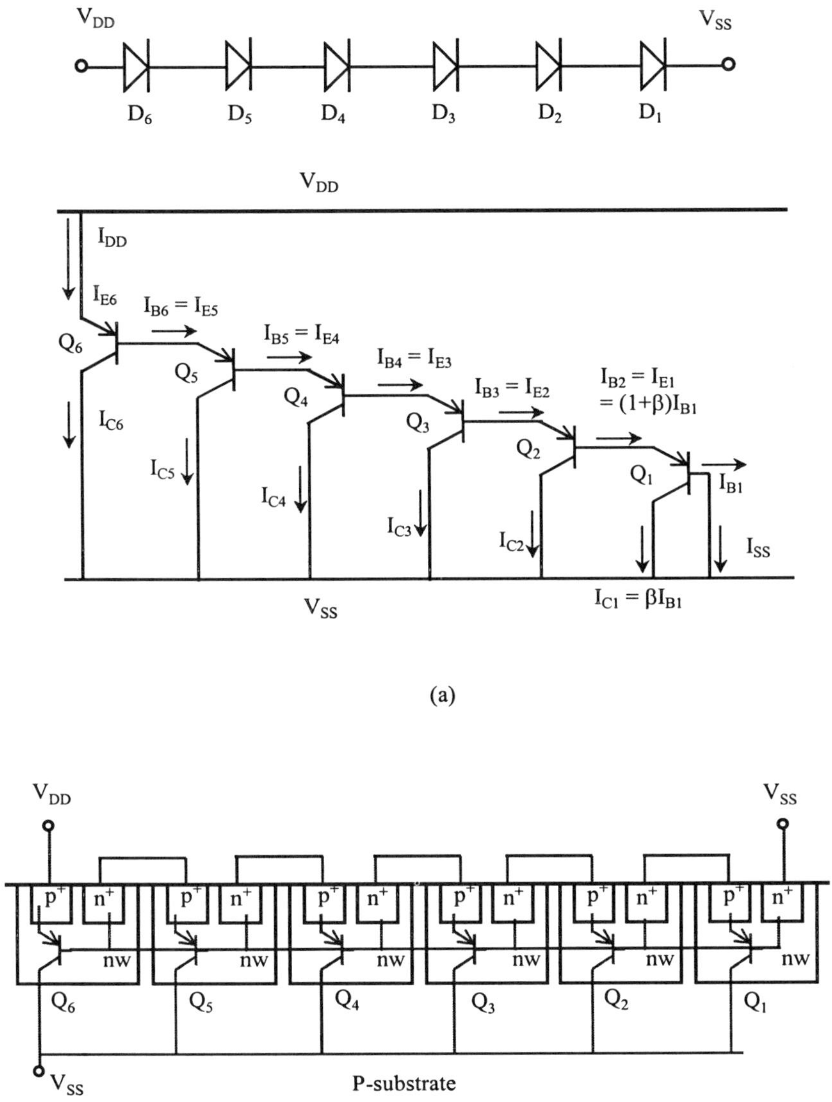

(b)

Figure 4-21 A six-PNP-diode chain power clamp: (a) schematics, (b) cross-section [34] (*Reproduced here by kind permission of Elsevier Science and authors*).

$$V_{D2} \approx nV_T \ln \frac{I_{E2}}{I_S} = nV_T \ln \frac{I_{E1}}{(1+\beta)I_S} = V_{D1} - nV_T \ln(1+\beta), \quad (4.4)$$

and

$$V_{D6} \approx nV_T \ln \frac{I_{E6}}{I_S} = V_{D1} - nV_T \ln(1+\beta)^5. \quad (4.5)$$

Assuming the same forward BE junction turn-on of V_D, the total trigger voltage is given by,

$$V_{t1} = \sum_{i=1}^{6} V_{Di} = 6V_D - 15nV_T \ln(1+\beta), \quad (4.6)$$

or generally, for an m-diode string, one obtains,

$$V_{t1} = \sum_{i=1}^{m} V_{Di} = mV_D - \frac{m(m-1)}{2} nV_T \ln(1+\beta). \quad (4.7)$$

This clearly indicates that the ideal linear V_D add-up feature collapses due to the substantial BJT gain β: the higher the β of a vertical PNP, the more reduction in the total turn-on voltage of the diode chain. Consequently, more diodes would be needed to realize a working power clamp for 3.3V applications. However, a longer PNP diode string may increase the discharging channel resistance, R_{on}, and leakage current significantly. The discharging resistance, R_{on}, looking into the V_{DD} pad, can be estimated using a BJT T-model in common-collector (CC) configuration, resulting an equivalent circuit shown in Figure 4-22. Starting leftward from Q_1 and assuming a zero-load at V_{SS} pad, simply exchanging the input and output resistances in the original CC amplifier formulas and progressing along the diode chain, the total R_{on} is approximated as,

$$R_{i1} \approx r_e + \frac{R_W}{1+\beta}, \quad (4.8)$$

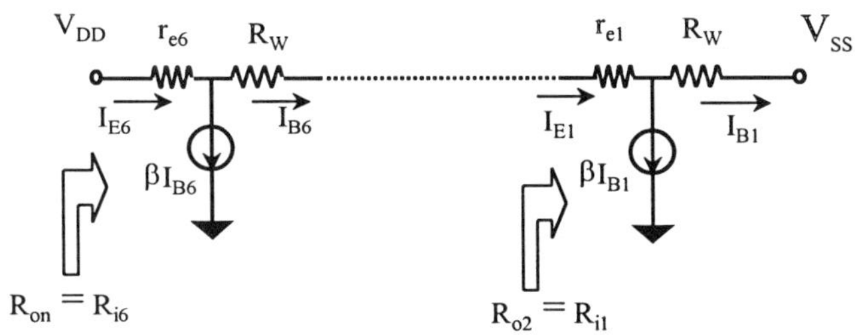

Figure 4-22 An equivalent circuit for the PNP diode chain in T-model.

and

$$R_{on} = R_{i6} \approx r_e + \frac{R_{i5} + R_W}{1+\beta} = r_e \sum_{a=0}^{5} \frac{1}{(1+\beta)^a} + R_W \sum_{b=1}^{6} \frac{1}{(1+\beta)^b}. \quad (4.9)$$

where r_e and R_W are series emitter resistance and parasitic lateral n-well (base of PNP) resistance, respectively. Without getting engaged with the complexity of mathematics, one can conclude that a higher β will reduce R_{on}. Now, back to the physical meaning, higher ESD performance requires lower discharging impedance, hence, better current handling capability. In this sense, a higher PNP β, or Darlington amplification, is beneficial to ESD protection. Of course, one should not conclude that the more diodes, the higher ESD robustness. Increasing the number of diodes in the chain will always lead to a larger R_{on}, though a higher β may reduce the magnitude of the increase trend. Next, let's consider the leakage current flowing into the V_{DD} port. From Figure 4-21, it is fairly obvious that any tiny leakage originated from the p-substrate/n-well junction in Q_1 will be amplified through the Darlington chain. One would expect a substantial leakage measured at the V_{DD} node. But, could one say any initiating leakage in Q_6 (i.e., the Q's in the upper chain) will be attenuated the β downward from V_{DD} to V_{SS}? This would be a good mind game for the readers. From the above discussions, it is clearly that a great deal of design trade-offs ought to be balanced in PNP diode chain design. Basically, the Darlington effect will reduce the final trigger voltage, increase leakage currents, while boost ESD protection capability as a side benefit. To make the situation worse, the substantial heat generation during ESD events leads to lattice temperature increase, which makes Darlington amplification even more significant. Generally, an IC designer concerns more about acceptable leakages and precise trigger voltage, the Darlington effect should therefore be suppressed. The basic means to crash the Darlington effect is to prevent the PNP Q's from getting into active mode. Based upon classic BJT transistor biasing techniques, voltage snubbing techniques across a few BE junctions, which clamps the V_{BE}'s to lower than the turn-on level, can achieve this goal. Implementation includes using a resistive voltage divider or a diode snubber as illustrated in Figure 4-23 [34]. It is worth to recall that, initially, diodes are preferred as a simple solution. However, to make it work as a power clamp, a diode chain becomes no longer a simple design as expected. As always, *should a designer start to think like a philosopher in chip design, the outcome might not be rosy.*

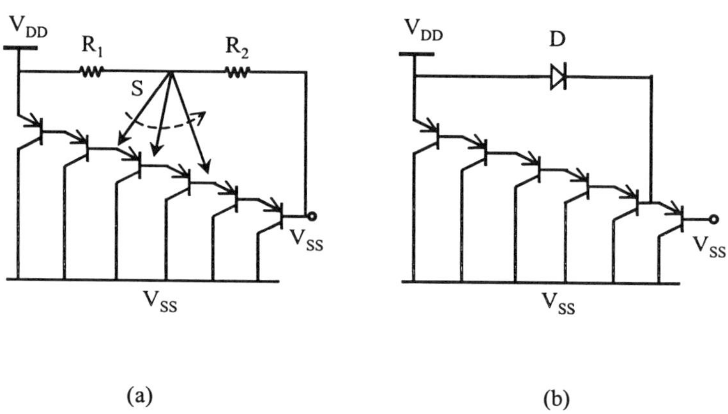

Figure 4-23 The Darlington effect of a PNP diode chain can be suppressed by: (a) using a resistive voltage divider, or, (b) using a snubber-diode.

4.3.4 Switch as Power Clamp

NMOS switches can work as good power clamps between power buses. The idea is to connect a sufficiently large NMOS device between a pair of power lines, e.g., V_{DD} and V_{SS}, which is controlled by a simple logic. The NMOS switch stays off in normal circuit operation. Under ESD pulses, the control logic produces a high signal that drives the NMOS into active turn-on mode, forms a low-impedance conducting channel and discharges ESD transients. Figure 4-24 shows one original implementation of a NMOS switch power clamp [36] that consists of a NMOS switch, M_1, a inverter chain of three, an R_1-C_1 coupling sub-net and a reset resistor, R_2. The time constant associated with the R_1-C_1 sub-net is designed to be much shorter than that of the circuit (typically in the ms range) to avoid interfering with circuit operation and fairly longer than that of interested ESD duration (e.g., 150ns for HBM ESD pulses) to ensure efficient coupling under ESD transients. As an ESD transient appears at a V_{DD} pad, the input node for the first inverter is pulled low due to R_1-C_1, a proper high signal is produced after the three inverters, and forces the M_1 turn on and operate in full active mode. A low-impedance discharging path is then formed. The large R_2 serves to reset the clamp after ESD transients. In addition to selecting the right R_1-C_1 sub-net, a sufficient large M_1 is necessary to ensure low discharging impedance in order to handle large ESD transients. The typical dimensions given in Figure 4-24 offers a good feeling about the practical sizes for such kind of a NMOS switch clamp. The advantage is that the clamp circuit can be simulated by IC designers using SPICE. However, the large size needed is one of the main drawbacks associated with such NMOS switches that cause chip layout, area budgeting and parasitic effect problems.

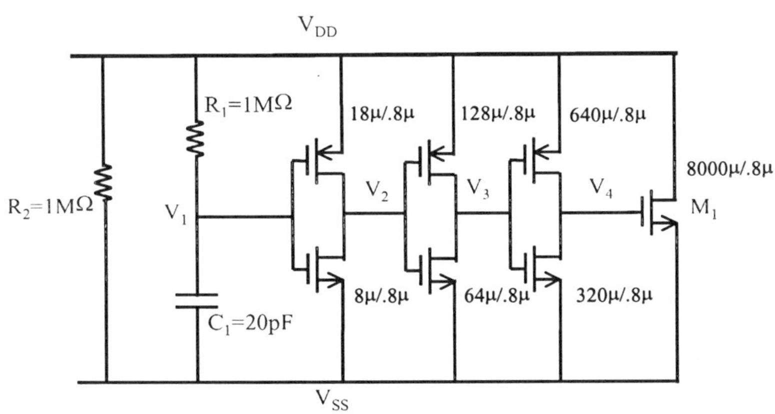

Figure 4-24 A NMOS switch power clamp schematic [36] (*Reproduced here by kind permission of the ESD Association and authors*).

4.4. SUMMARY

This chapter discusses circuit level ESD protection schemes. These ESD protection sub-circuits are developed to meet ever-increasing demands for higher ESD robustness as well as new technology features and circuit specifications. Multiple-finger and gate-coupled MOS structures are commonly used for input protection. For output pads, self-protection using output buffer transistor is an alternative to the dedicated ESD protection structures. Low-trigger SCR structures of various flavours become more attractive as chip size, parasitic effect, and layout become concerns in advanced applications, provided latch-up could be controlled. Power clamps using NMOS, SCR, diode chain and NMOS switch are used to protect power buses. Assuming a proper low-impedance discharging path and voltage clamping means being assured, any kind of ESD protection sub-circuits can be devised for special applications. However, one is cautioned to consider the net benefits before resorting to any complex ESD protection circuits. The penalties associated with the large sizes, increased parasitic effects, and layout problems should never be underestimated. Provided adequate ESD protection, a simple structure is always the better option. A few compact multiple-mode ESD protection structures meeting such rule will be discussed in the next Chapter 5.

REFERENCES

[1] Rountree, R. and Hutchins, C., "NMOS Protection Circuitry", *IEEE Trans. Elec. Dev., ED-32*, 1985, pp.910-917.

[2] Duvvury, C., Taylor, T., Lindgren, J., and Kumar, S., "Input Protection Design for Overall Chip Reliability", *Proc. EOS/ESD Symp.*, 1989, pp.190-198.

[3] Hulett, T., "On Chip Protection of High Density NMOS Devices", *Proc. EOS/ESD Symp.*, 1981, pp.90.

[4] Amerasekera, A. and Duvvury, C., *ESD in Silicon Integrated Circuits*, New York, Wiley, 1995.

[5] Scott, D., Giles, G., and Hall, J., "A Lumped Element Model for Simulation of ESD Failures in Silicided Devices", *Proc. EOS/ESD Symp.*, 1986, pp.41-47.

[6] Chen, K.-L., "Effect of Interconnect Process and Snapback Voltage on the ESD Failure Threshold of NMOS Transistors", *Proc. EOS/ESD Symp.*, 1988, pp.212-219.

[7] Polgreen, T. and Chatterjee, A., "Improving the ESD Failure Threshold of Silicided nMOS Output Transistors by Ensuring Uniform Current Flow", *Proc. EOS/ESD Symp.*, 1989, pp.167-174.

[8] Feng, H. G., "A Mixed-Mode Simulation-Design Methodology for on-Chip ESD Protection Design", *MS Thesis*, May, 2001.

[9] Kamata, T., Tanabashi, K., and Kobayashi, K, "Substrate Current due to Impact Ionization in MOSFET", *Jpn J. Appl. Phys., 15*, 1976, pp.1127.

[10] Sze, S. M., *Physics of Semiconductor Devices, 2nd Ed.*, Wiley, New York, 1981.

[11] Toriumi, A, "Experimental Study of Hot Carriers in Small Size Si-MOSFETs", *Solid State Electron., 32 (12)*, 1989, pp. 1519-1525.

[12] Tam, S., Ko, P., and Hu, C., "Lucky-Electron Model of Channel Electron Injection in MOSFET's", *IEEE Trans. Elec. Dev., ED-31 (9)*, 1984, pp.1116.

[13] Tam, S and Hu, C., "Hot-Electron-Induced Photon and Photocarrier Generation in Silicon MOSFET's", *IEEE Trans. Elec. Dev. ED-31 (9)*, 1984, pp. 1264.

[14] Fischetti, M, Laux, S. and Crabbe, E., "Understanding Hot-Electron Transport in Silicon Devices: is there a Short Cut?" *J. Appl. Phys.*, 1995, pp.1058.

[15] Chan, T. Y., Ko, P. K., and Hu, C., "A Simple Method to Characterize Substrate Current in MOSFETs", *IEEE Elec. Dev. Letts, EDL-5*, December 1984, pp.505.

[16] Hu, C., "Hot-Carrier Effects", Chapter 3, *Advanced MOS Device Physics*, Einspruch, N. G., and Gildenblat, G., Eds, Vol. 18, VLSI Electronics Microstructure Science, Academic Press, San Diego, CA, 1989, pp. 119-160.

[17] Duvvury, C. and Diaz, C., "Dynamic Gate-Coupled NMOS for Efficient Output ESD Protection", *Proc. Int. Rel. Phys. Symp.*, 1992, pp.141-150.

[18] Krakauer, D. and Mistry, K, "Circuit Interactions during Electrostatic Discharge", *Proc. EOS/ESD Symp.*, 1994, pp.113-119.

[19] Corsi, M., Nimmo, R. and Fattori, F., "ESD Protection of BiCMOS Integrated Circuits Which Need to Operate in the Harsh Environments of Automotive or Industrial", *Proc. EOS/ESD Symp.*, 1993, pp.209-213.

[20] Mack, W. and Meyer, R., "New ESD Protection Schemes for BiCMOS Processes with Application to Cellular Radio Designs", *Proc. IEEE, 0-7803-0593-0*, 1992, pp.2699-2702.

[21] Tandan, N., "ESD Trigger Circuit", *Proc. EOS/ESD Symp.*, 1994, pp.120-124.

[22] Smith, J., "A Substrate Triggered Lateral Bipolar Circuit for High Voltage Tolerant ESD Protection Applications", *Proc. EOS/ESD Symp.*, 1998, pp.63-71.

[23] Chatterjee, A. and Polgreen, T., "A Low-Voltage Triggering SCR for on-Chip ESD Protection at Output and Input Pads", *IEEE Elec. Dev. Letts., Vol. 12, No., 1*, January 1991, pp.21-22.

[24] Duvvury, C. and Amerasekera, A., "Advanced CMOS Protection Device Trigger Mechanisms During CDM", *Proc. EOS/ESD Symp.*, 1995, pp.162-174.

[25] Ker, M., Chang, H. and Wu, C., "A Gate-Coupled PTLSCR/NTLSCR ESD Protection Circuit for Deep-Submicron Low-Voltage CMOS IC's", *IEEE J. Solid-State Cir., Vol. 32, No. 1*, January 1997, pp.38-50.

[26] Chen, J. Amerasekera, A. and Vrotsos, T., "Bipolar SCR ESD Protection Circuit for High Speed Submicron Bipolar/BiCMOS Circuits", *IEEE IEDM Digest*, 1995, pp. 337-340.

[27] Ker, M., "Lateral SCR Devices with Low-Voltage High-Current Triggering Characteristics for Output ESD Protection in Submicron CMOS Technology", *IEEE Trans. Elec. Dev., Vol., 45, No. 4*, April 1998, pp.849-860.

[28] Carbajal, B., Cline, R. and Anderson, B, "A Successful HBM ESD Protection Circuit for Micron and Sub-Micron Level CMOS", *Proc. EOS/ESD Symp.*, 1992, pp.234-242.

[29] Voldman, S., Gross, V., Hargrove, M., Never, J., Slinkman, A., O'Boyle, M., Scott, S and Delecki, J., "Shallow Trench Isolation Double-Diode Electrostatic Discharge Circuit and Integration with DRAM Output Circuitry", *Proc. EOS/ESD Symp.*, 1992, pp.277-288.

[30] Ker, M., Wang, K., Joe M., Chu, Y. and Wu, T., "Area-Efficient CMOS Output Buffer with Enhanced High ESD Reliability for Deep Submicron CMOE ASIC", *Proc. IEEE Int. ASIC Conf.*, 1995, pp.123-126.

[31] Wu, C. and Ker, M., "ESD Protection for Output Pad with Well-Coupled Field-Oxide Device in 0.5μm CMOS Technology", *IEEE Trans. Elec. Dev., Vol. 44, No. 3*, March 1997, pp.503-505.

[32] Chan, T. and Culver, D, "ESD Protection Circuit", *U.S. Patent, No. 5,329,143*, 1994.

[33] Scott, D, Bosshart, P and Gallia, J, "Circuit to Improve Electrostatic Discharge Protection", *U.S. Patent, No. 5,019,888*, 1991.

[34] Voldman, S., Gerosa, G., Gross, V., Dickson, N., Furkey, S. and Slinkman, J., "Analysis of Snubber-Clamped Diode-String Mixed Voltage Interface ESD Protection Network for Advanced Microprocessors", *J. Electrostatics, Elsevier Science, 38*, 1996, pp.3-31.

[35] Dabral., S. and Maloney, T., "Designing on-Chip Power Supply Coupling Diodes for ESD Protection and Noise Immunity", *Proc. EOS/ESD Symp.*, 1993, pp.239-249.

[36] Merrill, R. and Issaq, E., "ESD Protection Methodology", *Proc. EOS/ESD Symp.*, 1993, pp.233-237.

Chapter 5

ADVANCED ESD PROTECTION

Mixed-Signal, RF and Whole-Chip ESD Protection

Successful ESD protection design ought to take into consideration two important factors: First, ESD protection is case-dependent, e.g., process technologies, chip layout and applications, all playing roles in ESD protection design. Particularly, ESD protection for mixed-signal and RF ICs poise new challenges to IC designers. Second, ESD protection design is not about designing individual ESD protection devices, rather, it is a whole-chip design task that demands for well-planned full-chip ESD protection schemes. This chapter discusses such advanced ESD protection topics.

5.1. ESD PROTECTION FOR MIXED-SIGNAL ICs

ESD protection design for mixed-signal IC chips requires special considerations in the design procedures if one wishes to achieve an optimised protection solution at full chip level [1, 2]. Several issues should be addressed in mixed-signal IC ESD protection design. The first unique feature in mixed-signal ICs stems from the fact that multiple power supplies, i.e., V_{DD} or V_{SS}, usually exist on a mixed-signal IC chip. It is not unusual to find a widely spread local power supply spectrum, anywhere from ±1V to ±100V. One issue associated with the multiple-supply chips is that one single type of ESD protection structure may not work for all pads across over a chip, because pad-specific customisation may be needed for optimal whole-chip protection. For example, an ESD protection structure of a trigger voltage V_{t1} = 5V is suitable for a V_{DD} = 3.3V circuit portion, while an ESD protection device of V_{t1} = 23V is good for a V_{DD} = 15V circuit section [3]. However, using either ESD protection structure globally may cause problem, i.e., short-circuit or slow turn-on of an ESD protection structure that may lead to early ESD failure. A good rule of thumb in designing proper triggering voltage is to set it higher than local V_{DD} with sufficient safe margins to avoid accidental turn-on of the ESD protection devices due to power bus fluctuations, while maintaining a low V_{t1} for easy triggering. The key point here is to achieve optimal full-chip ESD protection, which is often not possible with a one-for-all protection structure.

The second main issue in mixed-signal ESD protection design is associated with potential accidental turn-on of an ESD protection structure in mixed-signal operation due to substantial digital noises. A rough theory follows. It is recently observed in transmission-line pulse (TLP) testing that early triggering of ESD protection devices occurs as the rise time, t_r, of TLP pulses decreases, which is attributed to displacement current effect due to variations in the ramping ratios of TLP pulse rising edges, referred to as the dV/dt effect [4, 5]. Further discussion over the dV/dt effect will be given in the next section for RF ESD protection. While more research is needed to confirm the dV/dt theory, the V_{t1}~t_r phenomenon might cause accidental turn-on of ESD protection devices in mixed-signal operation, leading to chip malfunction. For example, it is reported that, in a mixed-signal CMOS transceiver chip, substantial digital noises were injected into the analog portion of the chip from the noisy digital circuit, resulting in noise signal with dV/dt of $\sim 1.15 \times 10^7$ V/s [6]. According to the observation and analysis to be discussed in the next section, such a high noise variation ratio may cause reduction in triggering of an ESD protection structure in non-ESD conditions. Hence, short-circuit malfunction might occur in mixed-signal ICs under normal operation. Proper design of mixed-signal ESD protection structures and special noise isolation techniques are therefore required in mixed-signal ESD protection design. Other issues may emerge in mixed-signal ESD protection design as well.

5.2. ESD PROTECTION FOR RF ICs

Driven by the amazing successes and exponential increase in demand of wireless communications, RF IC design becomes a red-hot field in semiconductor field. Consequently, a new challenge in ESD protection design emerges – ESD protection for RF ICs. However, while RF ESD becomes a popular topic in the filed, one ought to ask exactly what is unique in RF ESD protection design that makes it different, if any, from ESD protection for other applications. Unfortunately, RF ESD protection is currently still a research topic in its infancy that is under heavy investigation. No well-shaped theory exists yet for RF ESD protection. Nevertheless, it is fairly clear that several special issues must be addressed in RF ESD protection design. The first one is the parasitic effect of ESD protection structures on RF IC circuit performance and the second issue is associated with accidental triggering of ESD protection units by the high-frequency RF signals. It is fairly easy to understand that, while providing necessary ESD protection, an ESD protection structure inevitably produces parasitic effects – something fits into the "*no free lunch*" category very well. From chip design viewpoint, this is part of the problem defined as ESD-circuit interaction [2, 7-9], which depicts the mutual influences between the ESD protection structures and the core circuits being protected. Basically, full chip ESD protection performance may be limited by the core circuits being protected. For example, pre-mature ESD failure may occur due to parasitic discharging links within the core circuit. This is referred to as circuit-to-ESD influence. Meanwhile, ESD-induced parasitic effects may affect circuit performance as well, referred as ESD-to-circuit influences. Detailed discussions over such ESD-circuit interactions will be given in Chapter 9. The ESD-to-circuit influence becomes a real problem in RF IC design because high-frequency chips are normally very sensitive to parasitic parameters. Typical ESD-induced parasitic effects include RC delay due to parasitic resistance and capacitance of ESD protection structures, substrate noise coupling due to substantial parasitic ESD capacitances, as well as extra noise generation by ESD protection structures. Significant circuit performance degradation in both RF and mixed-signal IC chips due to ESD-induced parasitic effects were observed, ranging from clock corruption to noise performance to other general parameters, such as, slew rate, settling time and safe margins, etc [7-9]. Since RF ICs are extremely sensitive to the ESD-to-circuit influences, one critical requirement in RF ESD protection design is to devise novel low-parasitic compact protection structures. A few such low-parasitic ESD protection structures will be discussed in the next section.

Because of its high-frequency nature, RF signals might cause undesired accidental turn-on of ESD protection on an RF IC chip, leading to short-circuit problem. As mentioned previously, TLP test data show that there exists a clear relationship between the triggering voltage of ESD protection devices and the rise time of a TLP pulse [4, 5, 10, 11]. Figure 5.1 clearly shows this V_{t1}~t_r phenomenon obtained in TLP tests of various types of ESD protection structures, e.g., NMOS and SCR devices, with the rise time used in the measurements varying from 200ps to 20ns. Although the extent of the V_{t1}~t_r dependence varies among different ESD protection devices, with some following near linear curves and others showing saturation trend, the V_{t1}~t_r variation does exist. Such a V_{t1}~t_r relationship means that an ESD protection

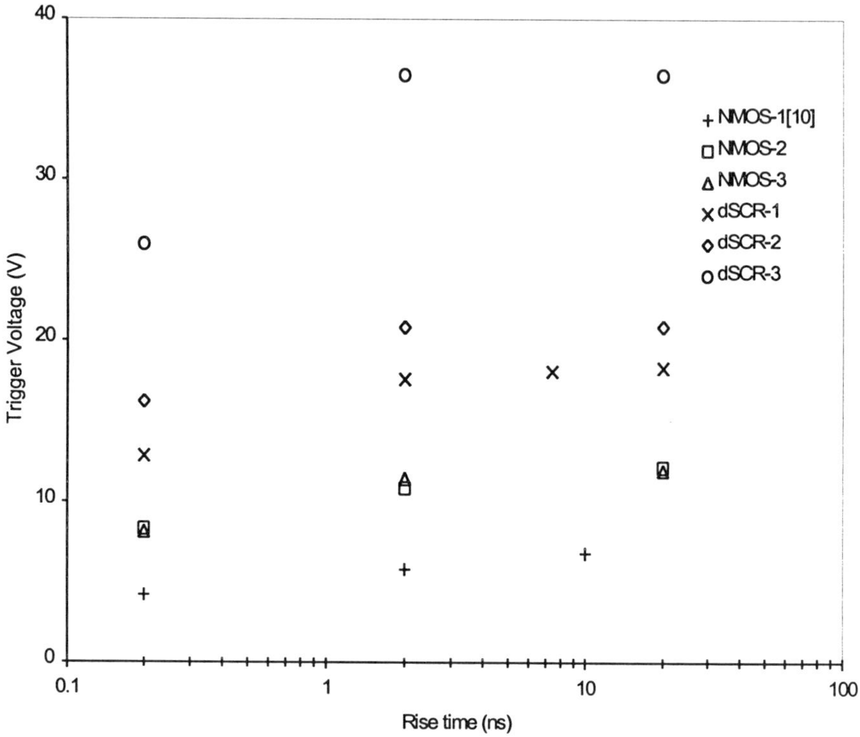

Figure 5.1 Typical ESD protection device trigger voltage versus rise time of the ESD pulses used from TLP measurements on a variety of ESD protection structures [11] show clear variation in trigger voltages, some in monolithic mode and others with saturation trends. The reduction in trigger voltage under shorter pulse rise time may lead to short-circuit problem in normal operation in mixed-signal and RF ICs.

structure may be turned on at different voltage levels depending upon the types of ESD pulses, instead of operating at a designed triggering voltage point.

In one theory, yet to be confirmed, the displacement current stemming from the substantial dV/dt rate of the rising edges of ESD pulses, i.e., varying rise time t_r, is believed to be the root cause to the V_{t1}~t_r

phenomenon. For example, in a ggNMOS ESD protection structure shown in Figure 5.2, the NMOS device works as a parasitic NPN BJT under ESD stressing because its drain junction is reverse-biased to avalanche breakdown and the hole current flows through a lateral parasitic resistor, R, builds up a voltage

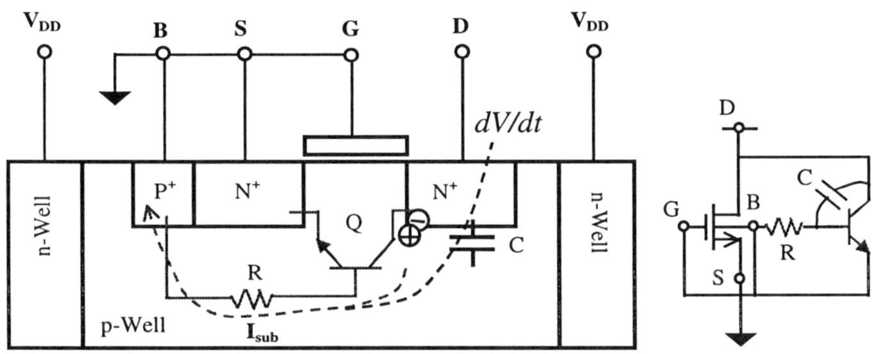

(a) (b)

Figure 5.2 A cross-section for a ggNMOS ESD protection device (a) and its equivalent circuit (b) show *dV/dt* displacement causing triggering even before avalanche breakdown occurs.

drop and forward turns on the source junction. Suppose there is a substantial dV/dt ratio associated with an incoming positive ESD pulse. Since the reverse-biased drain junction may has sufficient junction capacitance, C, a significant displacement current, given by $i = C\ dV/dt$, is produced upon ESD injection, which contributes to the substrate current that flows through the resistor R. Consequently, the ggNMOS device can be turned on even before avalanche breakdown occurs at the drain junction, leading to substantial reduction in ESD trigger voltage. Similarly, in an SCR ESD protection structure, as shown in Figure 5.3, its triggering mechanism is that the pwell/n-sub junction is reverse-biased to avalanche breakdown by an ESD pulse, the hole current flows through lateral parasitic pwell resistor, R, and builds up a forward bias of 0.65V at the N+/pwell junction, hence turns on the vertical NPN BJT Q2. If a significant displacement current generated by an ESD pulse is coupled into the pwell via the pwell/n-sub junction capacitance, C, the SCR may be triggered off as well, before avalanche breakdown occurs at the p-well to n-sub junction. The next question to ask is

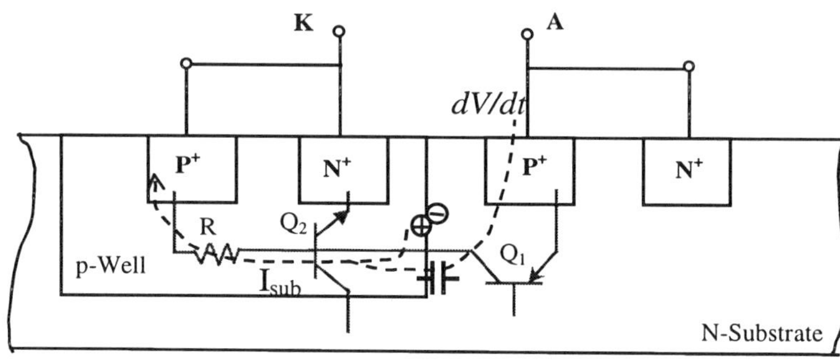

(a)

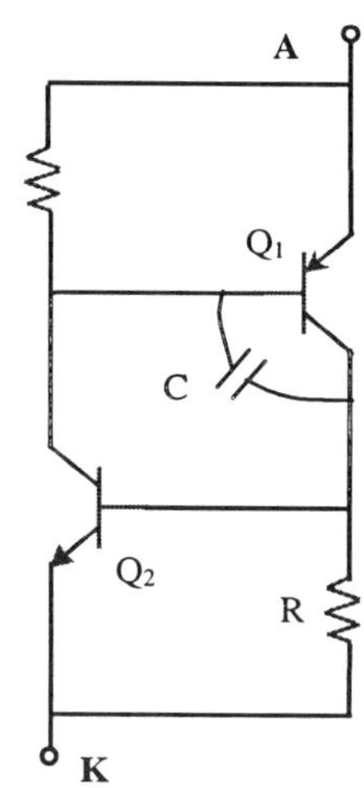

(b)

Figure 5.3 A cross-section for an SCR ESD protection device (a) and its circuit model (b) indicate *dV/dt* displacement current may lead to accidental turn-on of the SCR before avalanche breakdown occurs.

that how high a dV/dt ratio is needed to build up a forward PN junction bias, i.e., ~0.65V in silicon, in an ESD protection structure and whether such a desired dV/dt rate can be produced by an ESD pulse. For the first part of the question, a rough calculation was done for a set of NMOS and SCR ESD protection structure and the triggering threshold dV/dt ratios are found to be from 3×10^{10} V/s to 1×10^{11} V/s for the ESD protection structures measured [11]. These estimated data points for ESD protection structures are shown in Figure 5.4 as solid makers. For the second part of the question, a group of dV/dt data points obtained for a typical set of HBM ESD zapping testers, real HBM ESD waveforms and TLP testers gives a range from 7×10^{8} V/s to 1×10^{11} V/s as shown in Figure 5.4 in hollow makers [5]. Clearly, these two data sets roughly fall in the same bandwidth, though these data come from fairly limited sources. The significant meaning derived from Figure 5.4 is that it supports the dV/dt displacement theory in understanding the V_{t1}~t_r phenomenon in ESD protection design. Now back to the concerned RF ESD protection design to check whether an ESD protection structure may be accidentally triggered by an RF signal in normal operation. While recognizing the frequencies and magnitudes of RF signals vary widely and without attempting to cover all types of RF signals, a few dV/dt data points are derived from some *everyday* RF IC as well as mixed-signal IC papers, resulting in a dV/dt of ~2.5×10^{8} V/s for a 2.5GHz CMOS clock recovery circuit [12], ~4.3×10^{7} V/s for a 1GHz CMOS clock synthesizer [13] and ~1.23×10^{7} V/s caused by 7.1MHz digital clock noise coupling in a mixed-signal CMOS receiver chip [6]. These RF IC signal data, also shown in Figure 5.4, are clearly approaching the troublesome dV/dt bandwidth that starts to trigger the V_{t1}~t_r phenomenon. Hence, it is fairly natural to believe that, considering possibly stronger signals and higher frequencies, input and output RF signals might accidentally turn on ESD protection structures at I/O pads, leading to chip malfunction due to short-circuit problems. Therefore, the V_{t1}~t_r phenomenon associated with RF signals ought to be taken care of during RF ESD protection design. For example, special care must be excised during design to prevent the dV/dt displacement current from spreading into critical substrate regions, such as using proper guard-rings. Active research on RF ESD protection is currently taking off to address this new design challenge.

5.3. LOW-PARASITIC MULTIPLE-MODE SOLUTIONS

From previous discussions, it is evident that the ESD-to-circuit influence due to ESD-induced parasitic effect is an important design concern at whole-

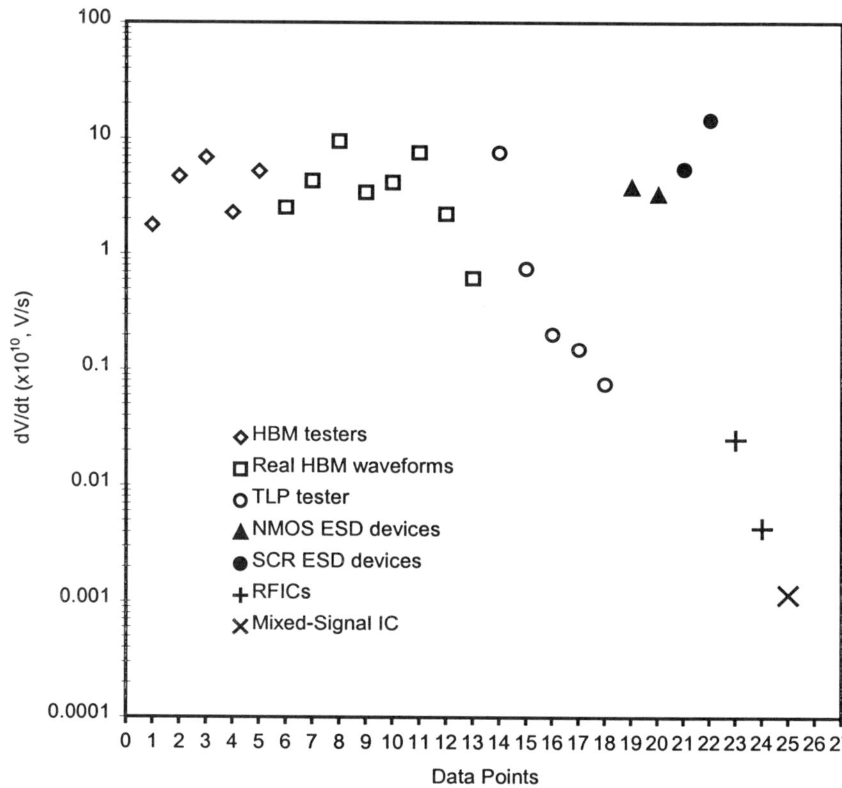

Figure 5.4 *dV/dt* data extracted from real HBM ESD waveforms, HBM zappers, TLP testers, and several ESD protection devices reside in the same neighbourhood, supporting the proposed *dV/dt* displacement current theory. Several data points for mixed-signal and RF ICs are very close to this *dV/dt* bandwidth, suggesting that accidental triggering of ESD protection structures might occur in non-ESD normal operation.

chip level. One of the most critical tasks in designing ESD protection structures for mixed-signal and RF ICs is to explore novel low-parasitic, compact ESD protection structures that bring in as less as possible parasitic effects as side products while maintain adequate ESD protection immunity. A few such innovative designs are discussed in this section.

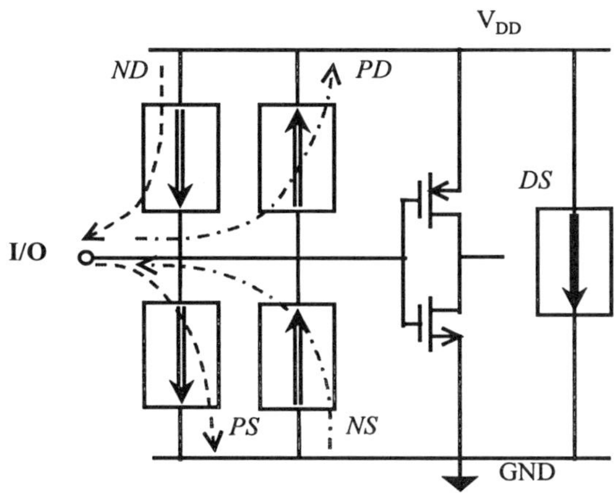

Figure 5.5 A complete ESD protection scheme using traditional one-direction ESD protection devices needs multiple ESD protection units for each pad.

5.3.1 A Dual-Direction ESD Protection Structure

In practice, ESD pulses applied to a bounding pad may come in different formats that may be classified as a PD mode for positive ESD pulse from I/O pad to V_{DD}, a ND mode for negative ESD transient from I/O pad to V_{DD}, a PS mode for positive ESD stress from I/O pad to V_{SS}, a NS mode for negative ESD pulse from I/O pad to V_{SS}, and a DS mode for positive ESD surge from V_{DD} to V_{SS} [14, 15]. These five ESD pulsing modes are illustrated in Figure 5.5. Well, what about an SD mode for positive ESD surge from V_{SS} to V_{DD}? This is a good question. Since all power line surges can be characterised as with respect to the global ground (GND) and one normally does worry much about any possible surge on earth, a DS mode seems have covered all power line surge situations. Of course, a positive surge from a negative local power bus, V_{SSx}, to a positive local power line, V_{DDy}, should be considered in whole-chip ESD protection design. With these ESD stressing definitions in hand, it is obviously that an ideal ESD protection scheme for any pad should be able to protect the pad against ESD transients of all the modes as defined in Figure 5.5. However, it is not unusual that an IC designer may not choose the complete ESD protection scheme in practical design because of technical and layout concerns, as well as availability of ESD protection structure for special circuits. At the time of this writing, one common situation is that little ESD protection, if any, is given for RF pads of IC chips in market. In Figure 5.5, separate ESD protection units are shown to safely discharge ESD pulses of different modes. This conceptual protection scheme has its ground because most traditional ESD protection structures, as introduced in previous chapters, operate in an asymmetrical mode, as shown in Figure 5.6a, which sets limitation on ESD robustness. For example, both NMOS and SCR ESD protection structure provide an active low-impedance discharging channel in one direction – normally being the goal of an ESD protection design task – while rely on the parasitic diode for ESD protection in the opposite direction.

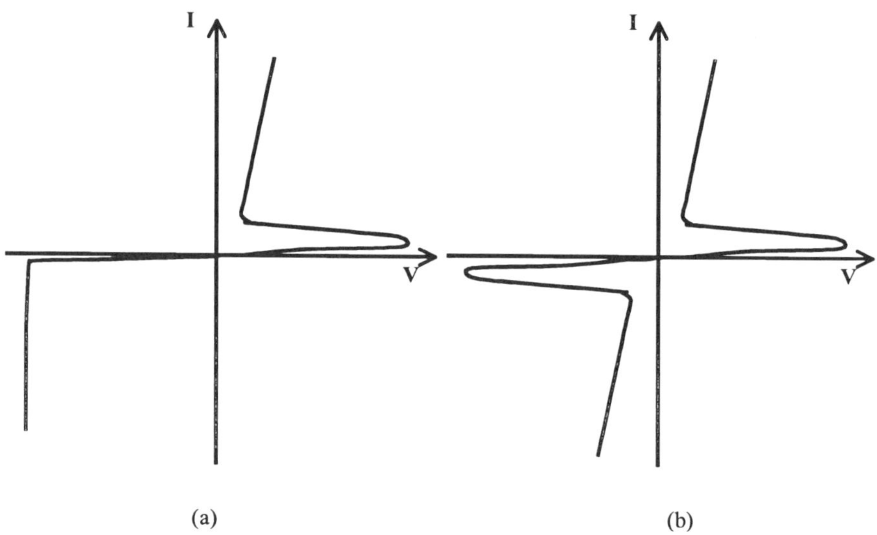

Figure 5.6 Typical I-V characteristics for ESD protection structure: (a) a one-direction asymmetrical operation, (b) a dual-direction symmetrical I-V curve.

Figure 5.7 shows a typical full ESD protection solution following this protection scheme, where four MOS-triggered SCR devices, as depicted in Figure 4.12, are used for each pad [16]. Though nothing is wrong in using parasitic structure for ESD protection, several issues arise. Firstly, any parasitic devices are not optimised for adequate ESD protection. The significantly higher discharging resistance associated a parasitic device substantially reduces the ESD protection level. In addition, the low forward turn-on voltage of about 0.65V in a diode prevents it from being used for a circuit with V_{DD} > 5V because of short-circuit concern. Therefore, a protection scheme shown in Figure 5.5 may be needed when using traditional ESD protection structures. The disadvantage of such protection scheme is obvious: too many devices, too much silicon consumption and too much ESD-induced parasitic effect.

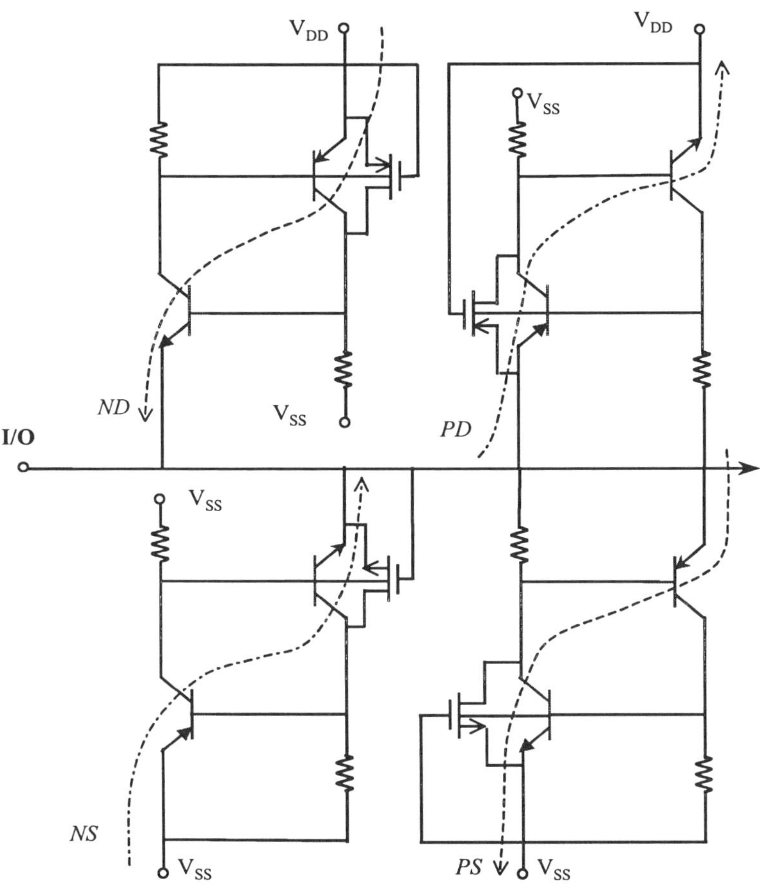

Figure 5.7 A four-device ESD protection scheme using one-direction MOSFET-triggered SCR units to achieve all active discharging for all four ESD stressing modes [16].

An innovative dual-direction ESD protection structure is given in Figure 5.8 [15], which depicts a two-terminal (anode A and cathode K) five-layer ($N_1P_2N_3P_4N_5$) device implemented in a BiCMOS technology, where an N-

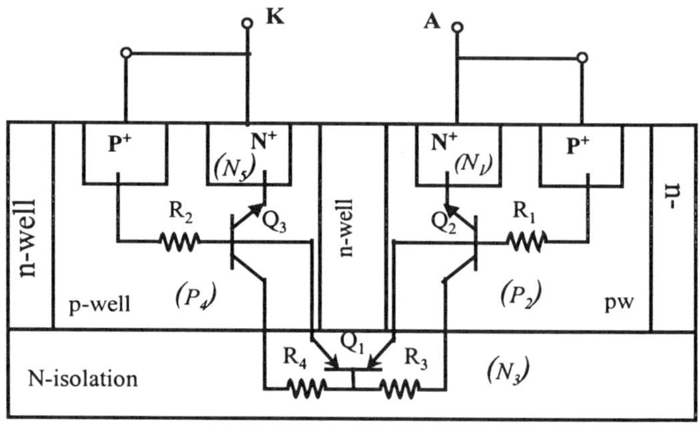

Figure 5.8 A novel dual-direction ESD protection structure in BiCMOS technology [15] (*Reproduced here by kind permission of IEEE and authors*).

type isolation layer was designed for noise decoupling. The corresponding intrinsic NPNPN structure and its equivalent circuit are shown in Figure 5.9. The whole structure consists of one lateral PNP transistor (Q_1=$P_2N_3P_4$), two vertical NPN transistors (Q_2=$N_1P_2N_3$ & Q_3=$N_3P_4N_5$) and parasitic resistors, R_1, R_2, R_3, & R_4. The ESD protection structure remains in off-state in normal operation. Under ESD stresses, these transistors work in pairs, i.e., Q_1-Q_3 in one direction and Q_1-Q_2 in the opposite direction, forming a reverse-connected SCR pair to protect internal circuits against ESD pulses in both directions, respectively. Considering a connecting case where the ESD protection device is connected between an I/O pad (terminal A) and ground (terminal K). Suppose a positive ESD pulse appears at the I/O pad with respect to GND (i.e., in PS mode), the BC junction (N_3P_4) of Q_1 is reverse-biased until reaching its avalanche breakdown, a large amount of electron-hole pairs are generated. Since the contact region P_4-P^+ is connected to the terminal K at GND, excess hole current will flow into terminal K, builds up a voltage drop over the parasitic R_2 and eventually forward turn on the BE (P_4N_5) junction of Q_3. The SCR device of Q_1-Q_3 ($P_2N_3P_4N_5$) starts to operate and forms an active low-impedance channel to safely discharge the ESD current transients and to clamp the I/O pad voltage to a sufficiently low level

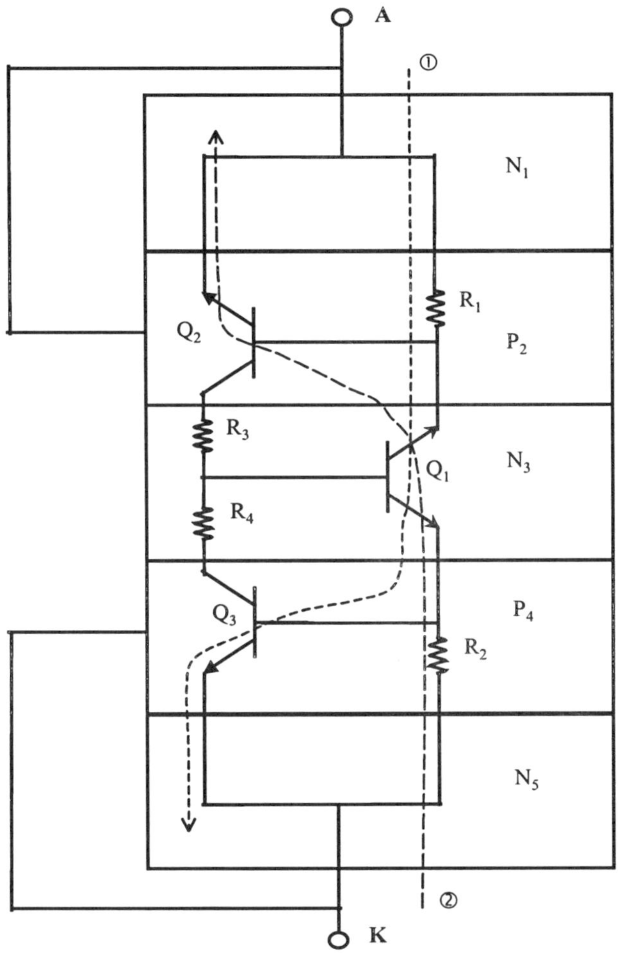

Figure 5.9 An equivalent circuit for the dual-direction ESD protection structure shows an intrinsic five-layer format device, which can discharge ESD transients in both directions, i.e., ① and ② [15] (*Reproduced here by kind permission of IEEE and authors*).

thus protecting ICs from being ESD-damaged. After the ESD pulse is over, the SCR discharges quickly and turns itself off as the current decreases to below its holding current level. The same mechanism applies to an NS mode

ESD event where the SCR of Q_1-Q_2 ($P_4N_3P_2N_1$) forms an active discharging path in the opposite direction. Therefore this new ESD protection device is capable of providing effective protection for ICs against ESD pulses in the two opposite directions, hence, forming an active low-impedance current discharging path in both directions, as depicted by the symmetric I-V characteristic in Figure 6b. Similarly, the new ESD protection unit connected from an I/O pin to V_{DD} will operate exactly the same way for the ND & PD ESD stressing modes. The ESD protection scheme using the new ESD protection structure is illustrated in Figure 5.10. The low holding voltage deep-snapback property indicates that the ESD protection device can handle large ESD current surges, translating into a high ESDV-to-Si ratio, where ESDV is defined as the ESD failure threshold voltage in kilo volts. Since the trigger voltage, V_{t1}, of the ESD protection device is determined by either junction avalanche breakdown or punch-through breakdown, the V_{t1} can be adjusted for different applications by either choosing different junction layers or varying the pwell-pwell spacing. External trigger-assisting sub-circuits, to be discussed in the next section, can also be used for deep reduction in the V_{t1}. The main advantage of this new dual-direction ESD protection structure is that it provides very high ESD protection level, features a high ESDV-to-Si ratio (~ 80V/μm width), only two ESD protection devices are needed for each pad to gain complete ESD protection

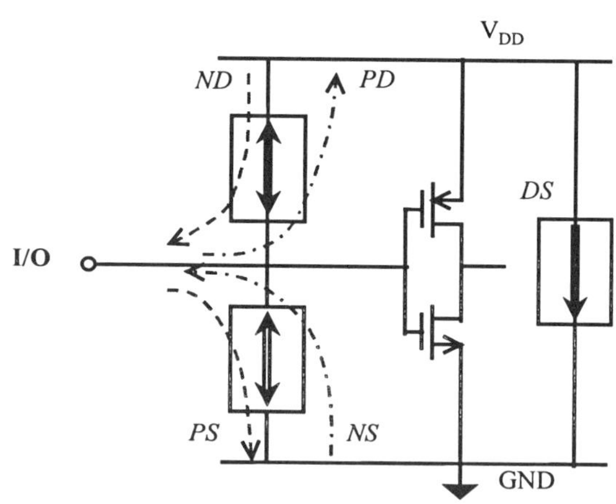

Figure 5.10 A complete ESD protection scheme using the dual-direction ESD protection structure.

as indicated by Figure 5.10, compared to the four-device solution shown in Figure 5.5. Consequently, Lower ESD-induced parasitic effects and less silicon consumption are expected when using the new ESD protection structure, which still ensures high ESD protection. This low-parasitic, compact ESD protection structure is particularly suitable for parasitic-sensitive RF ICs and area-sensitive high-pin-count chips. The wide spectrum of V_{t1}, ranging from 5V to 55V as reported [3], makes the new ESD protection structure a nice ESD protection solution for multiple-V_{t1} mixed-signal ICs as well. However, one ought to be cautious about the latch-up problem in design such kind of SCR-based ESD protection structures. One rule of the thumb in practical design is to ensure the holding current, I_h, be sufficiently higher than the highest on-chip operation current, often being the supply current, I_{DD}, in order to break the latch-up sustaining condition. In addition, proper layout is also critical in this matter. For example, placing an ESD protection device next to a huge output buffer transistor or a charge pump will guarantee a latch-up. The substantial substrate current coupled from these current-heavy devices may easily trigger the ESD protection structure accidentally and keeps it in the latch-up mode. Careful placement of ESD protection devices and use of guide rings may efficiently alleviate the problem [17]. Readers are referred to Chapter 7 for more discussions on layout issues.

5.3.2 An All-in-One Multiple-Mode ESD Protection Design

Referring to Figure 5.10, although the number of ESD protection devices used is reduced when using the new dual-direction ESD protection structure as compared with the scheme of Figure 5.5, one still has to use two protection units for each I/O pad plus one power clamp to realize complete ESD protection for the pad. Further reduction in protection device count will certainly beneficial to whole-chip design. Such an idea can be realized using an all-mode ESD protection structure as depicted in Figure 5.11 [18]. This new ESD protection structure can be viewed as a dual operation of the dual-direction protection device discussed above with shared centre diffusions as the anode terminal, A. The other two terminals are denoted as K1 and K2. This new three-terminal (A, K1 and K2) device contains three P-type wells, each has P^+ contact and N^+ contact diffusions. The structure consists of six bipolar transistors, Q_1, Q_2, Q_3, Q_4, Q_5, and Q_6, as well as parasitic resistors, which form two dual-direction SCR units, i.e., the dual-SCR of Q_1-Q_2-Q_3 and the dual-SCR of Q_4-Q_5-Q_6. For ESD protection, the three terminals, A, K_1, and K_2 of the ESD protection structure are connected to I/O pad, V_{DD} and GND, respectively, as illustrated in Figure 5.12. The equivalent circuit

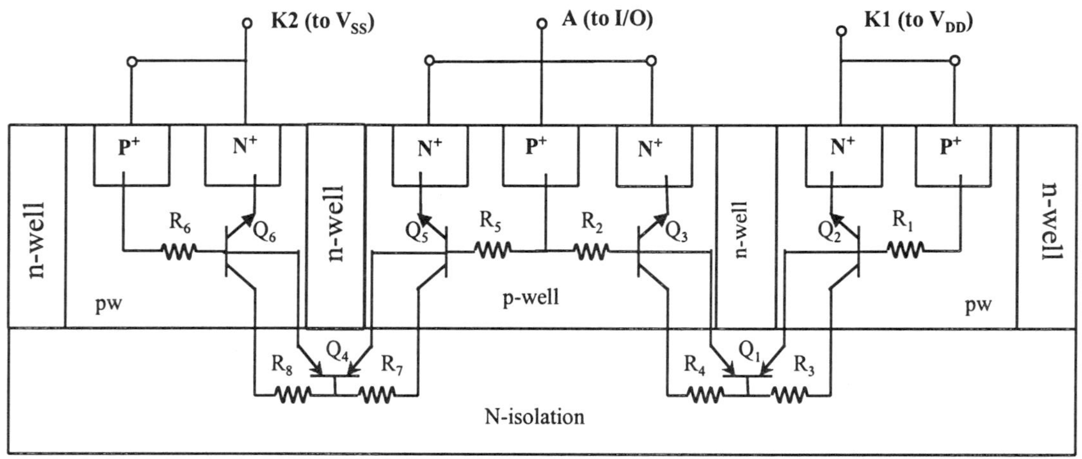

Figure 5.11 A cross-section for the novel all-in-one ESD protection structure [18] (*Reproduced here by kind permission of Elsevier Science and authors*).

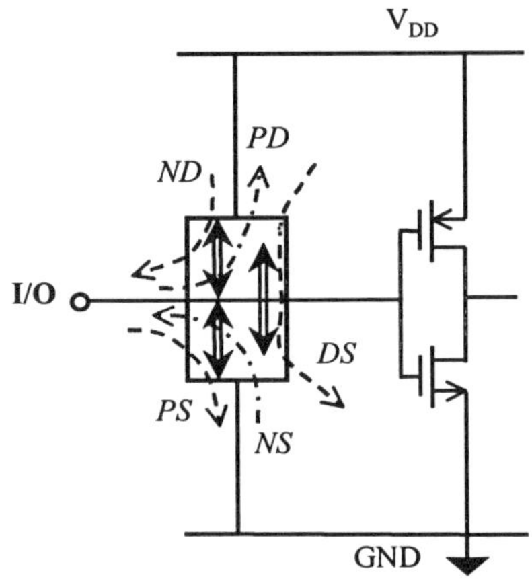

Figure 5.12 A complete ESD protection scheme using the all-in-one ESD protection structure is very area-efficient that reduces the ESD-to-circuit influences.

of the all-mode ESD protection structure is shown in Figure 5.13. Of course, the ESD protection structure must stay off in normal operation so as to not interfere with the core IC chip functionality. To describe its ESD protection operation, assume there is an ND ESD mode pulse occurs at the I/O pad, i.e., a negative ESD transient at terminal A with respect to V_{DD} (terminal K1). This is equivalent to applying a positive ESD pulse from terminal K1 to A. Similar to the operation in the dual-direction ESD protection device, the BC junction of Q_1 will reach to avalanche breakdown under the ESD stress and the substrate current will turn on Q_3, hence, triggers the SCR of Q_1-Q_3 and drives it into deep snapback mode with low holding voltage of $V_h \sim 2V$. An active shunting path, i.e., the ND as shown in Figure 5.13, with negligible impedance is formed to safely discharge the ESD current transient and to clamp the I/O pad voltage to a sufficiently low level, thus protects ICs from being damaged by a negative A-to-K1 ESD pulse. The same mechanism applies to the other four ESD pulsing modes: In the PD mode where a positive A-to-K1 ESD pulse occurs at I/O pad with respect to V_{DD}, the SCR of Q_1-Q_2 forms an active discharging path. In the PS and NS modes, a positive or negative A-to-K2 ESD pulse appears at the I/O pad, respectively, with respect to the ground. Correspondingly, the SCRs of Q_4-Q_6 or Q_4-Q_5 serves as the active discharging devices, respectively. In addition, if a DS mode ESD surge occurs at a power line, i.e., a V_{DD}-to-GND ESD pulse, the SCR of Q_7-Q_6 will provide ESD protection. Overall, only one such ESD protection structure is needed for each I/O pad to achieve full ESD protection, including the power clamping, hence named as an all-in-one ESD protection structure. The main advantage of this new all-mode ESD protection structure is that a single, very small device can provide complete ESD protection for each pad, therefore dramatically reduces the parasitic ESD effects and results in substantially less ESD-to-circuit influences [9]. This is a highly desirable feature for high-speed very-deep sub-micron IC

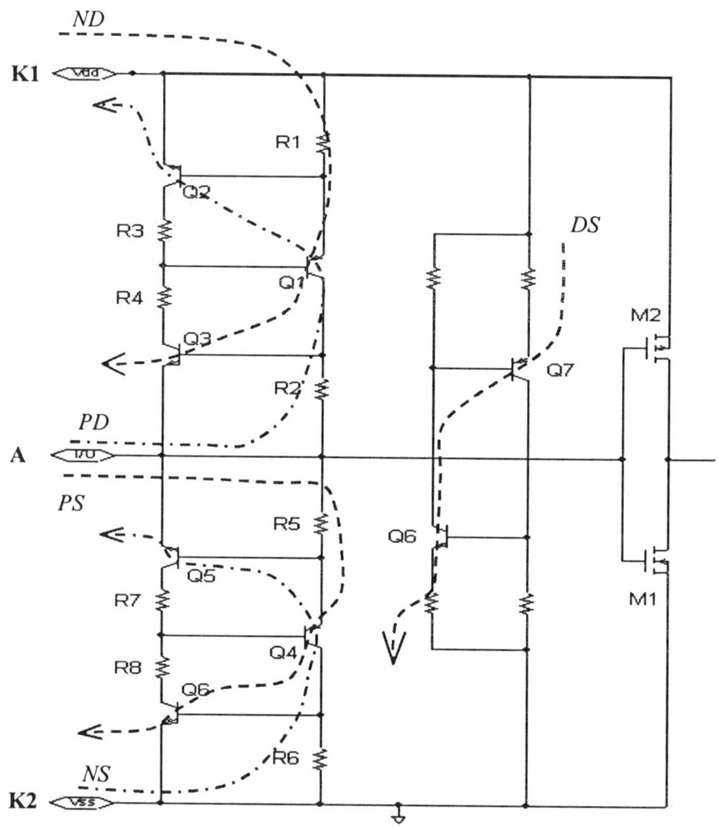

Figure 5.13 An equivalent circuit for the all-in-one ESD protection shows all five ESD discharging modes [18] (*Reproduced here by kind permission of Elsevier Science and authors*).

designs. Meanwhile, because of the reduced parasitic ESD capacitances, this new ESD protection structure features a shorter ESD response time that makes it a suitable solution for ultra fast ESD protection, e.g., IEC and CDM ESD tests [19, 20].

As mentioned before, an ESD protection device with adjustable trigger voltage is preferred in multiple-supply mixed-signal applications. The new all-in-one ESD protection structure offers tuneable trigger voltage that can be realized by choosing different diffusion layers as the avalanche breakdown junction, or, varying the well-to-well spacing for relatively lower punch-through breakdown. Alternatively, an external trigger-assisting sub-circuit cell may be used for deep reduction of the SCR triggering voltage. For example, a low-voltage all-mode ESD protection circuit is shown in Figure 5.14, where external current sources are used to reduce the triggering voltage, V_{t1}, and to adjust it to an appropriate level suitable to IC circuit sections with different local power supplies [3]. The idea is based upon the operation principle of the all-in-one ESD protection structure of Figure 5.11 discussed previously, where the V_{t1} is basically controlled by the efficient turn-on of the BE junctions of vertical NPN transistors (Q_2, Q_3, Q_5, and Q_6) upon breakdown of the Pwell/Nwell junction that is normally high. The mechanism of the new ESD protection circuit is that if some sort of external current can be injected into the parasitic well resistor that is in parallel with the NPN BE junction to accelerate the building-up of the V_{BE}, one should be able to reduce and control the trigger voltage. For example, in the PS mode ESD stress case, a switch (S3)-current source (I3)-resistor (R33) network is connected to the Q_4-Q_5-Q_6 section. The S3 is normally off and is turned-on by an ESD pulse. The I3 current flows through R33 that serves to build up the V_{BE} of Q_6 and eventually turns on this NPN transistor, therefore triggers the SCR of Q_4-Q_6. A variety of techniques may be used to form the S3-I3 network. As a simple example, a pair of back-to-back connected Zaner diodes, Dz1 & Dz2, as shown in Figure 5.14, is used to construct the S3-I3 unit. The Dz1 and Dz2 serve as the switch in the opposite biasing directions, respectively, to ensure the ESD protection circuit stays off in normal operation. The turn-on voltage of the Dz1 is carefully selected to control the trigger voltage of the ESD protection circuit. Similarly, Zener diode pairs can be used to form other S-I networks. Since a variety of Zener diodes are usually available in advanced BiCMOS technologies, one can easily tune up the ESD protection V_{t1} values to accommodate the special needs of different circuit sections with different local V_{DD} on a chip. It is therefore possible to design an optimised robust ESD protection circuit with small area and flexible trigger-voltage for mixed-signal and RF ICs.

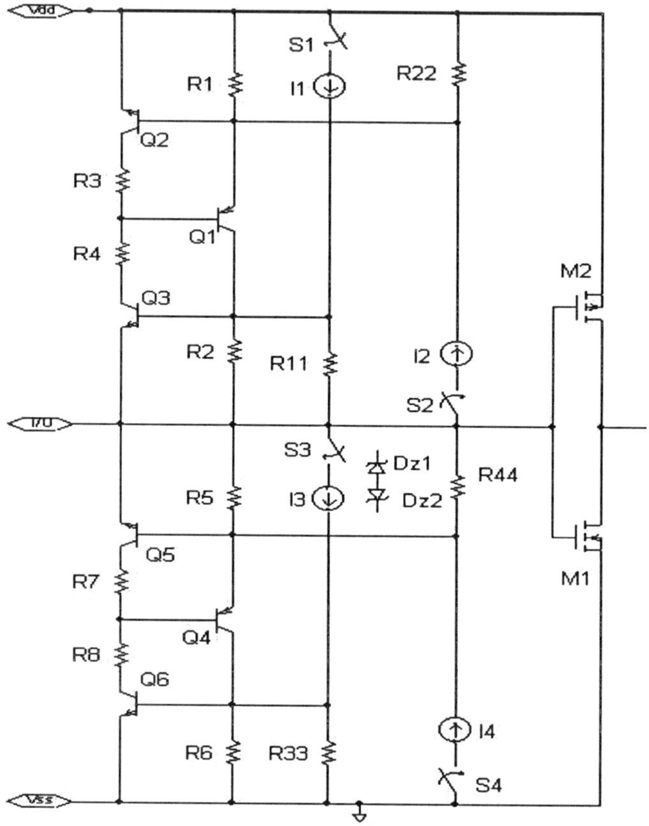

Figure 5.14 A low-trigger-voltage ESD protection circuit uses external trigger-assisting sub-circuit cells to realize adjustable low turn-on voltage [3] (*Reproduced here by kind permission of IEEE and authors*).

5.4. WHOLE-CHIP ESD PROTECTION SCHEMES

So far, readers should have had a good feeling that ESD protection design is not all about designing an individual high voltage high current handling device by your fellow workers in the process or device groups. Instead, ESD protection design is really a full chip circuit design task, well, at least if one wants to end up with an optimised chip performance. It is easier to design a functional ESD protection device, even a wonderful one, which, however, may not work at whole-chip level. It is not too hard to find an ESD protection solution for a few specific pads, while it takes significantly more mind work to realize a full-chip ESD protection. Finally, with a whole-chip ESD protection scheme in hand, one ought to take care of the ESD-to-circuit influences, as to be discussed in Chapter 9. Basic whole-chip ESD protection concepts are discussed in this section.

5.4.1 Principles for Full-Chip ESD Protection

Recall the discussions in Chapter 2 on typical ESD test models; almost all ESD test standards require ESD zapping stresses between IC pins in any combination in both directions. Therefore, to qualify for whole-chip ESD protection, there must exist a low-impedance conducting channel, preferably in active mode, from each pad to any other pads on a chip, which serves to safely discharge large ESD current transients as well as to clamp pad voltage to a sufficiently low level to avoid both thermal damage and dielectric rupture. The concept is fairly straightforward, however, a practical design may not be that easy. Out of all possible individual ESD protection structures published, many full-chip ESD protection schemes may be constructed [1, 21-23]. A few of them follow.

5.4.2 A Pad + Clamp Scheme

A typical whole-chip ESD protection scheme is illustrated in Figure 5.15 for a multiple-supply chip, which consists of I/O pad ESD protection plus power clamps, where all ESD protection units are shown as one-direction structures in clockwise/anti-clockwise connections. A little mind work will find that there does exist a discharging path between any two pads, regardless of the combination. For example, a PD mode ESD pulse from Pad 1 to V_{CC} will be absorbed by the ESD protection unit between Pad 1 and V_{CC1}, ESD1, conducting in active mode. In an ND mode ESD case at Pad 1, instead of relying on a parasitic conducting device in ESD1, a long discharging path formed by active ESD protection units, i.e., ESD 6-5-8-2, will take the role, assuming its combined impedance is much less than that of the reverse parasitic channel in ESD1. Similarly, a low-impedance active discharging path is formed for PS and NS mode ESD transients, respectively. Since multiple power supplies are used, a power clamp is placed between any two power buses to absorb power line surges in either DS or SD mode. In addition, a shunting channel exists between any two I/O pads, e.g., the Pad 1 and Pad 2, namely, ESD 1-6-5-9-4 for the positive pulse

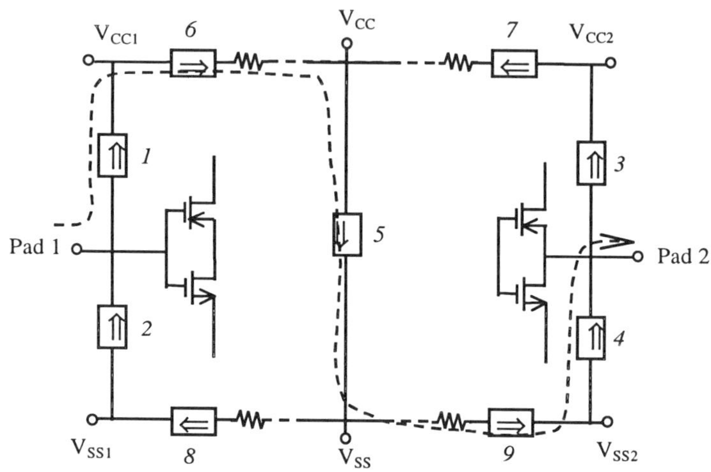

Figure 5.15 A pad-clamp whole-chip ESD protection scheme forms an active discharging channel between any two pads on a chip.

and ESD 3-7-5-8-2 for the negative transient, respectively. One critical checking point in designing such a whole-chip ESD protection circuit is that one ought to look into the total discharging channel impedance in long paths, at least the worst case that is the longest one between Pad 1 and Pad 2 as shown in Figure 5.15. This worst-case discharging resistance must be low enough in order for it to handle the given ESD current transients and the add-up clamping voltage through the ESD device chain must be sufficient low as well. Particularly, one can no longer neglect the metal power bus resistance in multiple-mode ESD current situations. Basically, all the individual ESD protection structures described in Chapters 3 and 4 can be used as ESD protection units in Figure 5.15. For example, using a simple reverse-connected diode for I/O pads and using a NMOS switching clamp for power line protection. If dual-direction ESD protection structures are available, such as that depicted in Figure 5.8, higher ESD robustness can be expected because all discharging channels, between any two nodes, will be short and active conducting paths with low impedance. One would not have to worry about the worst-case discharging problem as mentioned above. Of course, using a multiple-mode ESD protection structure, such as that shown in Figure 5.11, a much simpler whole-chip ESD protection scheme can be realized that can still maintain adequate ESD protection. In practical design, different ESD protection structures may be used at different location with diversified sub-circuit functionalities existing on a chip, typically in mixed-signal applications, for optimised full-chip performance.

5.4.3 A Common ESD Discharge Bus Scheme

In addition to the pad-clamp-oriented ESD protection scheme, a common ESD discharge bus can be used for full-chip ESD protection [1, 22]. As shown in Figure 5.16, this whole-chip ESD protection scheme consists of a global ESD discharge bus and dual-direction ESD protection units at each pad, including I/O pads and power supplies [1]. Since a dual-direction ESD protection device is placed at each pad, there always exists a low-impedance active shunting path between any two pads, with the common ESD discharge bus serving as a global conduction link. To ensure low resistance in the common ESD discharge bus, it needs to be wide enough. A simple estimation based upon given interconnect electronic design rule data helps to

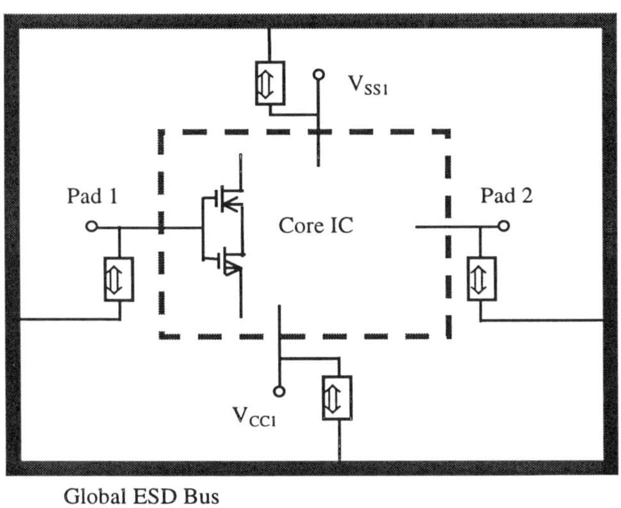

Figure 5.16 A whole-chip ESD protection scheme uses a global ESD discharging bus and dual-direction ESD protection devices [22] (*Reproduced here by kind permission of IEEE and authors*).

determine the metal line width for the ESD discharge bus. It is recommended to place a highly-doped diffusion line underneath and connected to the ESD discharge bus in order to improve its thermal dissipation capability, because silicon substrate is the only heat sink. Since dual-direction ESD protection structures are used, the common ESD discharge bus scheme results in lower discharging impedance, as compared to the pad-clamp solution discussed above, because there are always only two ESD protection devices in the conducting channel. It is also possible to use traditional single-direction ESD protection devices in the whole-chip ESD protection scheme, such as an NMOS, which has a parasitic diode operating in the reverse stressing case [22]. However, its current handling capability and application range may be limited. Another main benefit in using common ESD discharging bus is that fewer ESD protection devices may be needed, which translates into less ESD-induced parasitic effects and less ESD-to-circuit influences. This is a highly desirable feature for modern IC chips.

Numerous other design solutions are reported to realize full-chip and partial chip ESD protection. This is where an IC circuit designer can utilize his or her expertises as a chip designer, something one normally does not expect from his fellow device engineers.

5.5. NON-PORTABILITY IN ESD PROTECTION

"*We need ESD robustness of 2kV HBM for the new bluetooth radio transceiver chip in 0.35μm BiCMOS RF technology. How do we do it?*" asked the design manager. "*It should not be a problem. The memory group had a successful 4kV ESD protection structure for their chips in 0.25μm CMOS technology. We might just use that nice design.*" Answered the leading IC designer. Oops, here comes the common misunderstanding in ESD protection design: many designers believe that a successful ESD protection structure is portable, at least within the same process technology. The bloody reality is that, as in this bluetooth RF chip design case, the nice interoperability does not exist, at least not always. First of all, it is almost sure that ESD protection design is non-portable among different process technologies because of variations in key process parameters, such as, doping profiles, differences in substrates and epitaxy layers, etc. The kind of CMOS scaling concepts deeply rooted in designers' brains stop working in ESD protection design. Secondly, ESD protection design is often not portable even within the same process technology because ESD protection is extremely application-specific, as mentioned before. Taking the above conversation as an example, the successful ESD protection for digital chips may not work in analog applications. A 2.4GHz RF chip poises a significant challenge for ESD protection design because the high-frequency RF signal might lead to accidental triggering of ESD protection structures as discussed in Section 5.2. At least, the mixed-signal nature of the RF radio chip may lead to non-optimised ESD protection performance in terms of ESD-circuit interactions if using the digital ESD protection solution. Therefore, it is extremely important for IC designers to understand that there is no portability in modern ESD protection circuit design. Of course, it is possible to develop a nice ESD protection cell library for a specific process technology. However, since ESD protection design normally lags in the technology development chain and an IC process technology has limited lifespan, there has rarely been any successful story in using ESD protection cell library heard in commercial ESD protection design practices so far.

5.6. SUMMARY

In this chapter, several important advanced ESD protection design concepts are discussed. Modern mixed-signal and RF ICs require wide-angle considerations in ESD protection circuit design. The ESD-circuit interaction becomes an inevitable and critical issue in chip-level ESD protection design. The ESD-to-circuit influence must be considered in chip designs. Circuit malfunction due to possible accidental triggering of ESD protection structures caused by super fast RF signals emerges as a real problem. Although partial ESD protection is still acceptable in current practices, ESD protection design really desires a full-chip consideration. Basic ESD protection concepts, such as creating conducting paths, low-impedance discharging and low-voltage clamping are the basis for whole-chip ESD protection design. Finally, ESD protection design is typically application-specific. No portability shall be assumed in ESD protection circuit design even within the same process technology.

REFERENCES

[1] Feng, H. G., "A Mixed-Mode Simulation-Design Methodology for on-Chip ESD Protection Design", *MS Thesis*, Illinois Institute of Technology, May 2001.

[2] Wang, A. Z., Feng, H. G., Gong, K., Zhan, R. Y. and Stine, J., "On-Chip ESD Protection Design for Integrated Circuits: an Overview for IC Designers", *J. Microelectronics, Elsevier Science, Vol. 32/9,* August 2001, pp.733-747.

[3] Feng, H. G., Gong, K. and Wang, A., "An ESD Protection Circuit for Mixed-Signal ICs", *Proc. IEEE Custom IC Conf.*, May 2001, pp.493-496.

[4] Wang, A., "An Investigation into the V_{t1}~t_r Relationship in TLP Testing", a proposal submitted to the ESD Association, October 12, 2000.

[5] Barth, J. and Richner, J., "Correlation Consideration: Real HBM to TLP and HBM Testers", *Proc. EOS/ESD Symp.,* 2001, pp.453-460.

[6] Xu, M., Su, D., Shaeffer, D., Lee, T. and Wooley, A., "Measuring and Modelling the Effects of Substrate Noise on the LNA for a CMOS GPS Receiver", *IEEE J. Solid-State Circuits, Vol. 36, No. 3,* March 2001, pp. 473-485.

[7] Feng, H. G., Gong, K. and Wang, A., "ESD Protection Design Using Copper Interconnects: More Robustness and Less Parasitics", *SRC Publication: 2000 Publications in Copper Design Challenge, Pub. P000375,* 17-Mar-2000.

[8] Feng, H. G., Gong, K. and Wang, A., "A Comparison Study of ESD Protection for RFIC's: Performance vs. Parasitics", *Proc. IEEE RFIC Symp.,* 2000, pp235-238.

[9] Gong, K., Feng, H. G., Zhan, R. Y. and Wang, A. Z., "ESD-Induced Circuit Performance Degradation in RFICs", in press, *Microelectronics Reliability, Elsevier Science,* Oct. 2001.

[10] Barth, J., Verhaege, K., Henry, L., and Richner, J., "TLP Calibration, Correlation, Standards, and New Techniques", *Proc. EOS/ESD Symp.,* 2000, pp.85-96.

[11] ESD design database, The Integrated Electronics Laboratory, Illinois Institute of Technology.

[12] Anand, S. and Razavi, B., "A CMOS Clock Recovery Circuit for 2.5 Gb/s NRZ Data", *IEEE J. Solid-State Circuits, Vol. 36, No. 3,* March 2001, pp. 432-439.

[13] Foley, D. and Flynn, M., "CMOS DLL-Based 2V 3.2ps Jitter 1GHz Clock Synthesizer and Temperature-Compensated Tunable Oscillator", *IEEE J. Solid-State Circuits, Vol. 36, No. 3,* March 2001, pp. 417-423.

[14] Ker, M. D., Wu, C. Y., Cheng, T. and Chang, H. H., "Capacitor-Couple ESD Protection Circuit for Deep-Submicron Low-Voltage CMOS ASIC", *IEEE Trans. VLSI Systems, Vol. 4*, 1996, pp. 307-321.

[15] Wang, A. Z. and Tsay, C., "On a Dual-Polarity Electrostatic Discharge Protection Structure", *IEEE Trans. Elec. Dev., Vol. 48, No. 5,* May 2001, pp.978-984.

[16] Ker, M. and Wu, M., "CMOS on-Chip Four-LVTSCR ESD Protection Scheme", *U.S. Patent, No. 5,572,394*, Nov. 5, 1996.

[17] Wang, A., Tsay, C., Bielawski, J. and DeClue, L., "Design Optimisation of a Practical ESD Protection Circuit by CAD: A Case Study", *Proc. IEEE UGIM Symp.*, 1999, pp. 116-119.

[18] Feng, H. G., Gong, K. and Wang, A. Z., "A Novel on-Chip Electrostatic Discharge Protection Design for RFIC's", *J. Microelectronics, Vol 32/3, Elsevier Science*, March 2001, pp 189-195.

[19] *IEC 1000-4-2*, "Electromagnetic Compatibility, Part 4: Testing and Measurement Techniques, Section 2: Electrostatic Discharge Immunity Test", The International Electrotechnical Commission, January 1995.

[20] *ESD STM5.3.1-1999 Revised*, Electrostatic Discharge Sensitivity Testing – Charged Device Model Component Level, the ESD Association, 1999

[21] Ker, M., Wu, C., Wu, M., Yu, T. and Wang, A., "Whole-Chip ESD Protection for CMOS VLSI/ULSI with Multiple Power Pins", *IRW Final Report*, 1994, pp.124-128.

[22] Narita, K., Horiguchi, Y., Fujii, T. and Nakamura, K., "A Novel on-Chip Electrostatic Discharge Protection for Beyond 500MHz DRAM", *IEDM Tech. Digest,* 1995, pp.539-542.

[23] Eaton, J. Jr. and Horner, R, "A System Engineering Approach to the Design of on-Chip Electrostatic Discharge Protection", *IRW Final Report,* 1995, pp.22-28.

Chapter 6

ESD FAILURE ANALYSIS AND MODELING

ESD failure mechanism and signature may be the most argued topics in ESD protection design, particularly when dealing with data obtained by using somewhat less reputable MM or CDM ESD test models, as well as data for latent ESD damages. ESD testing equipment factors are among the big contributors to the confusions. However, understanding ESD failure mechanism is critical to successes of ESD protection circuit design. Though ESD failure analysis (FA) is not the centre focus of this book, typical failure analysis results and ESD failure models will be discussed in this chapter, aiming to offer IC circuit designers a ballpark picture of ESD failure mechanisms and design debugging techniques.

6.1. WHY ESD FAILURE ANALYSIS?

"*Why should I bother to look into ESD failure analysis when things seem to be uncertain about it anyway?*" This would probably be the kind of question raised by an IC designer while reading this chapter. Well, please just be a little bit open-minded and patient. There might be some quite healthy stuff in ESD FA work that would benefit IC designers in design practices. Indeed, things seem to in chaos in ESD failure analysis, at least to IC designers who are used to those well-defined iron-rules like the Kirchoff Laws. Every paper seems to report a different result that might only be valid for *that* device in *that* process technology, tested in *that* ESD model and done by using *that* piece of equipment. Results are often different across different ESD test models, e.g., HBM, MM, CDM, IEC, or, TLP, etc, for different process technologies, e.g., 1.5μm, 1.0μm, 0.8μm, 0.5μm, 0.35μm, 0.25μm, 0.18μm, or, 0.13μm. It is not uncommon to find controversial data reported in publications. However, FA work is very helpful in ESD protection circuit design debugging and may well be the way to ensure design success in developing new ESD protection solutions. It would be wise for an IC designer to take the fuzzy-logic thinking approach in dealing with FA results. In the final phase of IC production, ESD zapping test will be performed to quantify ESD robustness of every IC part by simply reporting whether one chip passes or fails a specific zapping level, hence stamping a "kV" number on the data sheet. However, such ESD zapping test results provide no insights into the root causes of a failed design. With CAD simulation techniques still being unable to fully guide ESD protection circuit design, or at least being not skilfully utilized to do so, ESD FA remains the most valuable debugging technique in ESD protection circuit design. A few interesting FA examples will be discussed in this chapter to demonstrate how one may benefit from FA results in improving ESD protection design. Interested readers are referred to references [1] provided in this chapter for in-depth discussions.

6.2. ESD FA TECHNIQUES

A variety of techniques and equipment can be used for ESD FA analysis, both electrically and physically. As discussed in Chapter 2, ESD qualification is done by performing ESD zapping tests for IC parts, where pin electrical characteristics are monitored against acceptable chip specifications during the pass-fail classification tests. Hence, electrical measurement, often conducted by ATE (automatic test equipment) means in an industrial environment, is the basis for FA analysis. Electrical tests can generally detect ESD damages of IC chips. Detailed ESD damage phenomena are then inspected by physical examination techniques, from which the findings are collected, analysed and classified. As a result, typical ESD failure signatures may be defined, ESD damage mechanisms will be understood, and valid correlation between physical damages and electrical characteristics are established. Therefore, useful information will be obtained to improve existing ESD protection circuit design and to direct future design activities.

A useful ESD FA action flow consists of visual inspection, electrical testing and physical inspection. ESD damages include dead-hard failures due to open or short circuit, hard failures with significant leakage current, and soft failures with noticeable but within-specification leakage current. Visual inspection is done by simply looking at wafers under microscopes that can usually easily tell dead-hard ESD damages caused by dielectric or metal blowout, i.e. burnout. In electrical evaluation, electrical failure criteria are defined for chip functionalities and IC parts are then electrically characterized against these parameters. If any pin parameter exceeds the relative criterion, a hard failure occurs. Otherwise, a soft damage may happen if noticeable deterioration is observed for any significant electrical parameter, which however is still within the given specification, and a lifetime problem may be expected. One commonly used electrical parameter is the pin leakage current, which can readily be measured using a curve tracer, a semiconductor parametric analyser, or a more sophisticated IC ATE testing system. Both hard and soft ESD damages can be detected and evaluated by leakage current measurements. It is worth to point it out that leakage current is not the only defect indicator. Theoretically, an IC chip should be evaluated for deviation from all electrical specifications specified on the data sheet for qualification. While electrical evaluation serves to make the final judgement on the level of chip damage in terms of functionality, it normally provides no insights into the defect root causes and failure mechanisms, which can be investigated by physical inspection. Once electrical inspection results in a failure in functionality, IC dies will be de-processed, or prepared in other means, and physical examination will follow to locate the defects and to inspect the damaged areas. Physical inspection, normally using fairly sophisticated inspecting equipment, provides multiple-dimensional morphological and/or atomic information with high resolution for de-processed wafers at both surface and sub-surface levels that enables one to pin-down defects and to study the failure signatures.

A variety of visual and physical inspecting instrumentation are available for ESD FA investigation. The simplest, but most efficient, tool is optical microscope that allows easy detection of both dead-hard burnout damages on the surface and sub-surface hard damages after de-layering. The main limitation of an optical microscope is its limited magnification, normally below 1,500X with resolution to 1μm or so. Much more sophisticated microscopy tools that offer very high resolution include scanning electron microscope (SEM), transmission electron microscope (TEM), atomic force microscope (AFM), scanning force microscope (SFM), liquid crystal analysis, light emission microscope, etc [2]. Another useful contactless electrical test technique is the electron beam test system (E-beam).

A SEM system has very high magnification (~150,000X) and resolution capability. IC dies are de-processed layer by layer and inspected by a SEM until a visible defect is spotted. A SEM system basically does surface analysis, however, limited three-dimensional image and atomic information can be obtained by using sample tilting skill and back-scattered electron technique, respectively. TEM technique has very high magnification that allows a resolution better than two angstroms in three-dimensional micro-structural analysis. Its disadvantage is its extremely involving sample preparation and highly destructive analysis nature. An AFM system uses one surface of a stylus to scan over another surface, i.e., the sample, and detect the atomic forces between the two surfaces. The mapping of force variation while scanning the sample is recorded for topographical and sectional analysis. Since ESD failures are usually accompanied by heat generation or photon emission, the liquid crystal and light emission analysis techniques are very useful in ESD FA studies. In liquid crystal analysis, the wafer is placed on a controlled heat chuck and covered by some type of liquid crystals. The chip is then powered up; a thermally-coloured image is taken where hot spots, i.e., defects, show as dark-coloured locations. A light emission microscopy technique is also referred as photon emission microscopy technique (EMMI), which generates a spectral image by detecting photons emitted from the powered IC dies due to currents, avalanche electron-hole generation, thermal radiation, tunnelling current through dielectric, and inter-band transitions, etc [3-7]. It emerges as a very powerful ESD FA analysis tool because ESD failures often result in high currents, heat generation, junction breakdown, and dielectric breakdown, where a large amount of photons are produced in the failure process. EMMI technique can pin-down ESD defect of 0.5 μm scale. More attractively, these emission techniques can be used in full-biasing conditions and allows one to investigate ESD failure evolution procedures as well as ESD damages under different operation conditions. Of course, these sophisticated techniques do not come for free. High cost-of-ownership, complexity in sample preparation and highly skilled operation are typical prices one ought to pay for.

6.3. SOME ESD FAILURE SIGNATURES

ESD damages in IC parts can be characterised as catastrophic failure, i.e., hard failure, and latent failure, i.e., soft failure. A catastrophic damage causes immediate malfunction of IC parts, reflected by some electrical parameters exceeding the given chip specifications. Typical hard failures are associated with either thermal damages in metal interconnects, contacts/vias, and silicon, or dielectric ruptures. Thermal damages are caused by high-current-induced Joule-heating, localized over-heating and heat distribution. Dielectric failure is typically caused by high electric field density under high voltage stresses, where gate oxide breakdown in CMOS transistors is the typical failure signature. There exists fairly good understanding of failure mechanisms for most hard ESD damages. Latent ESD failure is normally defined as the kind of ESD damages, often invisible, which lead to noticeable, however, still within the specification, electrical parameter deterioration after lower-level ESD stresses. An IC part suffering latent ESD damages still functions for the time being. In fact, most soft damages are recoverable upon thermal annealing. However, latent damage results in time-dependent performance degradation and reduces the lifetime of the IC parts. Typical latent ESD damages are associated with dielectric integrity problems, such as, time-dependent dielectric breakdown (TDDB) of gate oxide layers, and accompanied by leakage current increases. Though, exact latent ESD failure mechanism is still a topic subjected to further research currently. It is therefore very helpful to understand common ESD failure signatures in ESD protection circuit design practices. The following discussions provide such useful ESD FA information.

Taking the classic primary-secondary ESD protection structure depicted in Figure 4.9 in Chapter 4 as the first example, a real-world implementation for I/O pad protection is shown in Figure 6.1, where the primary ESD protection device (ESD_p) is a grounded-gate NMOS device (ggNMOS), the secondary ESD protection (ESD_s) is a metal gate thick-gate (FOX) NMOS device (Thick-NMOS), and a diffusion resistor, R, serves as an isolator [8]. ESD zapping tests using HBM, MM and CDM test models were excised for this ESD protection structure and ESD damage images for de-processed silicon are shown in Figure 6.2. In Figure 6.2a, HBM ESD damages are shown occurring at both ends of the Thick-NMOS device, featuring thermal filament from the contacts at drain end to the gate region and spreading into the source side. This is typically referred as a D-S silicon filament defect. Figure 6.2b shows similar D-S filament damages in a MM ESD zapped structure. In addition to the defects at the two ends, extra filament defect are observed across the drain region, which is attributed to the oscillatory nature of MM waveforms. Since these MM pulse ripples are fairly strong and have long duration (~30ns), each oscillatory MM ripple can cause similar D-S filament damage at different locations. Figure 6.2c shows the same type of D-S filament damage in CDM ESD zapping case, however, often less severe than the HBM damages. All right, how do this FA result help an IC designer? A little comparison found that the two end regions always suffer ESD damages (i.e., HBM, MM or CDM zapping), which is attributed to non-uniform current flows at the ends due to layout discontinuity (Figure 6.1b) as well as large thermal gradient at the boundary of heated shallow drain diffusions and the outside. This observation indicates that the end contacts could be the weakest points. Two improved layout designs, as shown in Figure 6.3, were therefore proposed to address the end-damage problem.

Again, it is important to realize that these specific solutions might very well not work for any other designs.

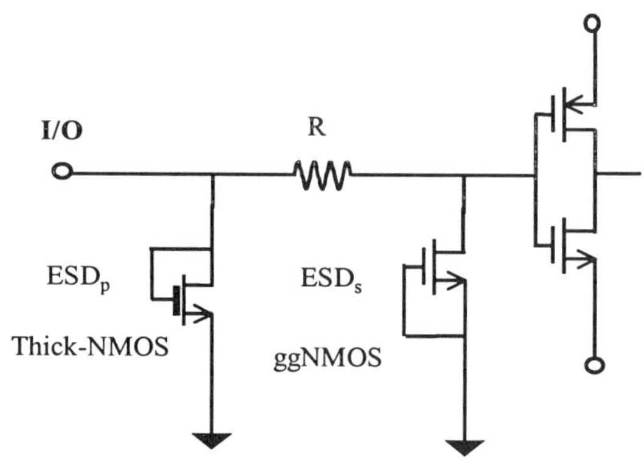

(a)

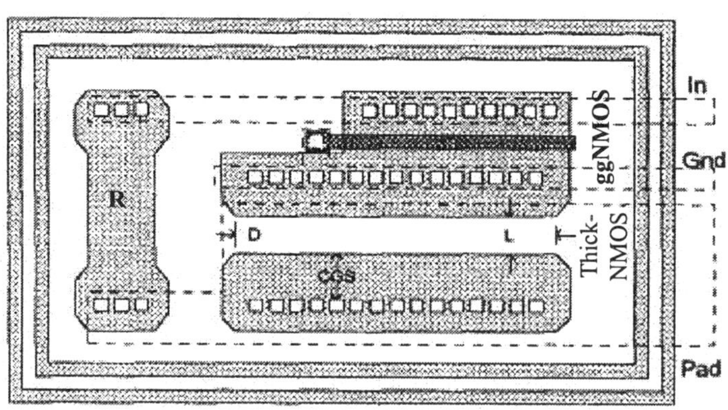

(b)

Figure 6.1 A primary-secondary ESD protection circuit: (a) schematic, (b) layout [8] (*Reproduced here by kind permission of the ESD Association and authors*).

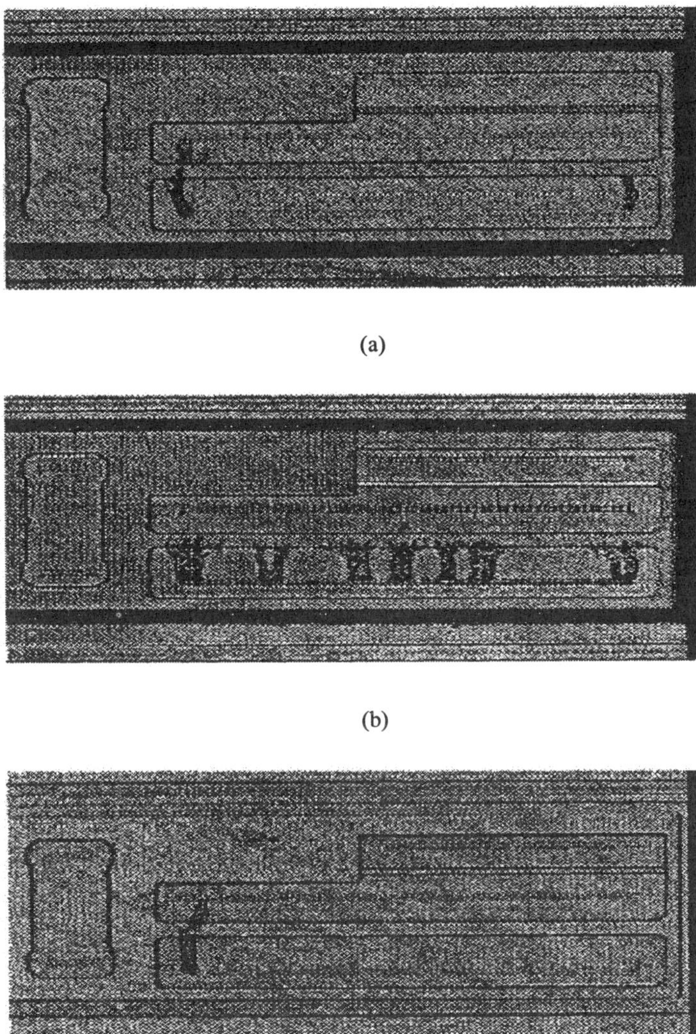

(a)

(b)

(c)

Figure 6.2 ESD damage images under ESD zapping for the primary-secondary ESD protection structure shown in Figure 6.1 under (a) HBM, (b) MM, (c) CDM stresses [8] (*Reproduced here by kind permission of the ESD Association and authors*).

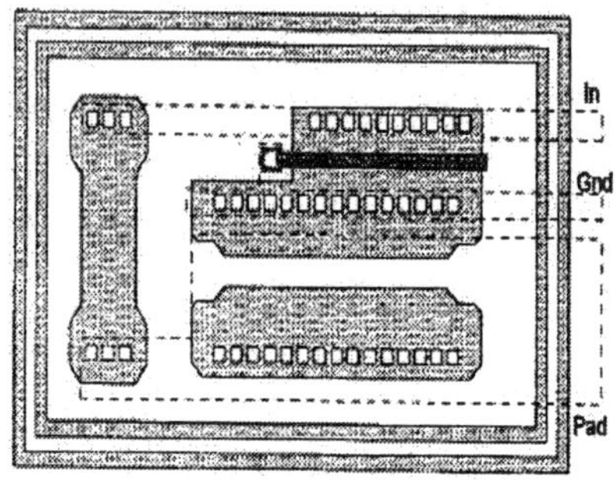

(a)

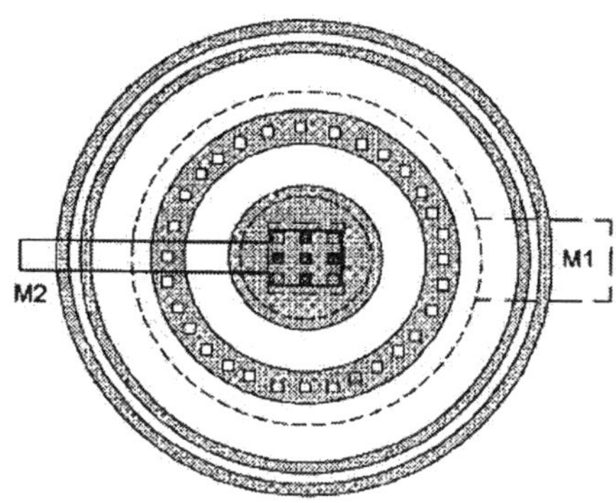

(b)

Figure 6.3 Improved layout designs for the Thick-NMOS to avoid the end-ESD-damage problem: from (a) to (b) [8] (*Reproduced here by kind permission of the ESD Association and authors*).

With adequate ESD protection, an IC chip shall be immune to ESD damages. However, ESD protection structures often do not work the way they are expected, consequently, internal circuit damages may occur. Very often, a parasitic device inside a circuit being protected may be turned on and forms a parasitic discharging path that will compete with the designed ESD protection structure. Since such internal parasitic device is never optimised for high-current handling capability, ESD damages may occur inside a protected circuit. Figure 6.4 shows such an internal failure case, where a HBM ESD-induced D-S Si filament was observed in an NMOS transistor of an input buffer circuit in a 0.35μm retrograde n-well salicided CMOS technology [9]. This type of D-S Si filament ESD defect is also a common failure signature in MOS ESD protection devices that are regularly observed in both HBM and CDM ESD stressing cases [10, 11]. Its failure mechanism is well understood as being associated with large ESD current transient flowing through a triggered lateral NPN bipolar transistor as explained in Chapter 3. Another classic ESD defect signature is the source or

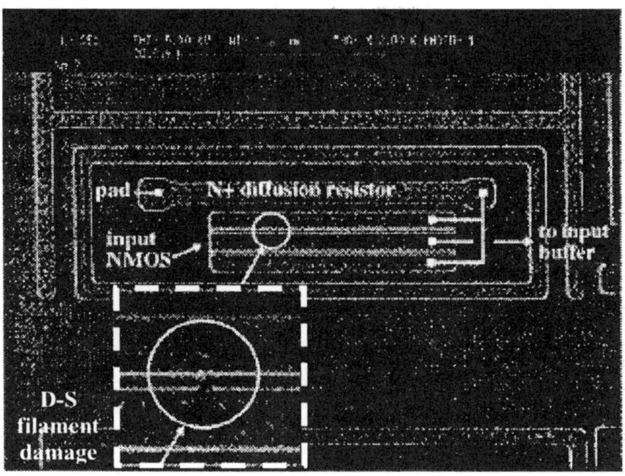

Figure 6.4 An example of internal circuit ESD damage: D-S Si filament defect in an input buffer NMOS transistor. The inset is a close-up image under high magnification [9] (*Reproduced here by kind permission of the ESD Association and authors*).

drain contact damage. Figure 6.5 illustrates these two types of NMOS ESD damages due to CDM zapping in a gcNMOS ESD protection circuit implemented in a 0.35μm CMOS technology [11]. The terminal-to-terminal Si filament ESD failure signature is also common in other ESD protection structures, for example, in ggNMOS-triggered SCR ESD protection structures similar to that depicted in Figure 4.11 in Chapter 4. Figure 6.6 shows examples of such ESD damages in low-triggering SCR ESD protection designed in 0.35μm CMOS process as well [11], where the difference is that a D-S Si filament (Figure 6.6b) occurs under positive CDM

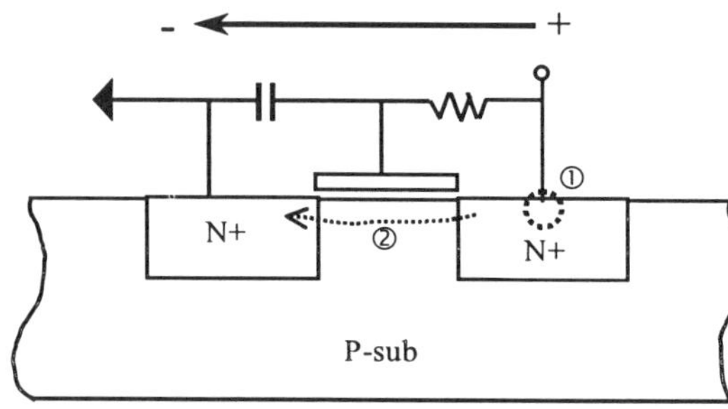

(a)

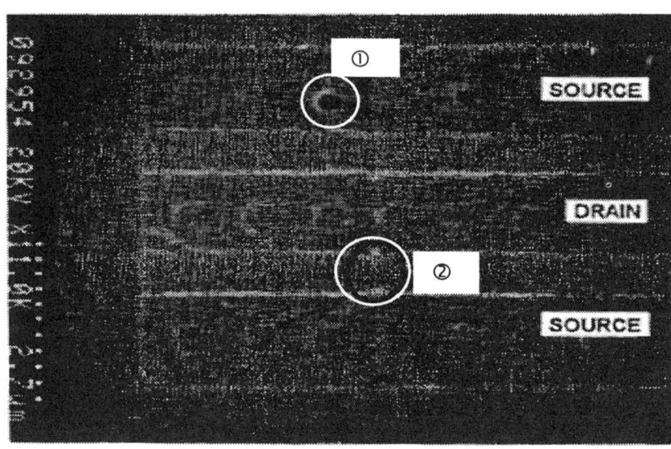

(b)

Figure 6.5 D/S contact (type ①) and D-S Si filament (type ②) CDM ESD damages in a gcNMOS ESD protection structure: (a) cross-section, (b) FA image [11]. These two ESD failure signatures are typical to MOS ESD protection structures that also occur in HBM stressing. (*Reproduced here by kind permission of the ESD Association and authors.*)

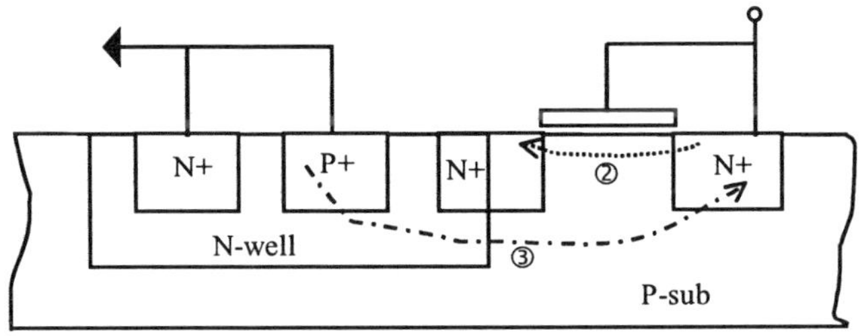

(a)

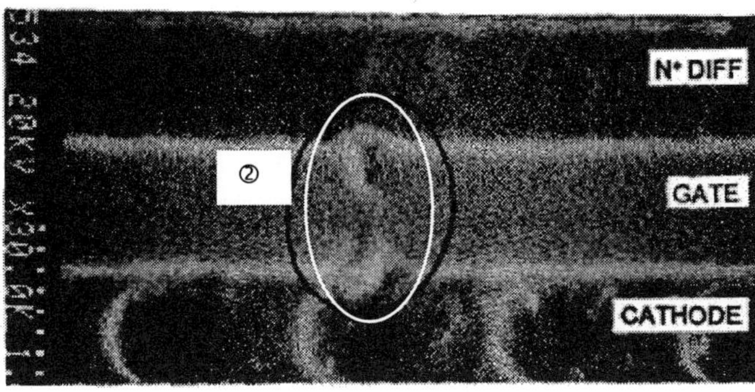

(b)

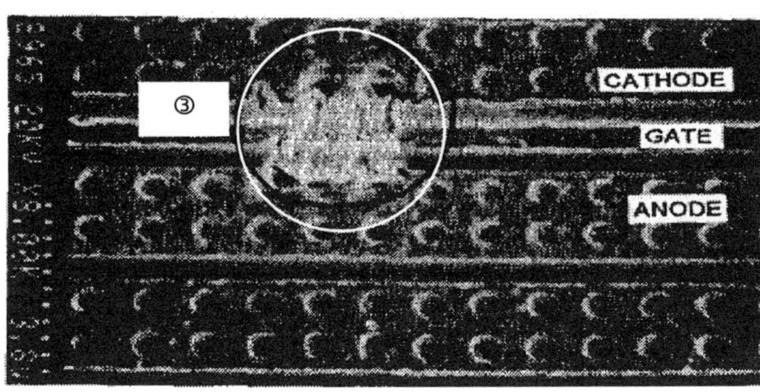

(c)

Figure 6.6 Typical terminal-to-terminal Si filament ESD damages in a ggNMOS-triggered SCR ESD protection: (a) cross-section, (b) D-S filament (type ②) in ggNMOS under positive CDM zapping, (c) Anode-Cathode filament (type ③) in SCR under negative CDM stressing [11] (*Reproduced here by kind permission of the ESD Association and authors*).

ESD zapping because the ggNMOS device conducts currents, while an anode-cathode Si filament (Figure 6.6c) happens in negative ESD stressing due to SCR operation. The cross-sectional structure in Figure 6.6a explains the ESD failure mechanism clearly. A clearer SEM image, obtained after deprocessing, for contact spiking ESD damage is given in Figure 6.7 for an Al/Si melt-through in a CMOS gate array circuit [12].

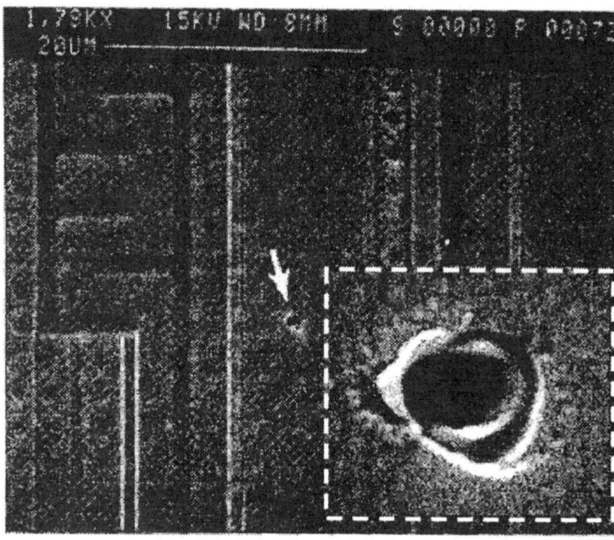

Figure 6.7 A contact spiking ESD failure signature in a CMOS gate array circuit [12] (*Reproduced here by kind permission of the ESD Association and authors*).

ESD damage in gate oxide films is another typical ESD failure signature that happens in all ESD stressing modes, e.g., HBM, MM and CDM. Oxide defects can appear at different locations and in different shapes for the same IC part depending upon ESD zapping modes. For example, Figure 6.8 shows gate oxide defects in bus interfaces of a data communication circuit in 1.5μm CMOS technology under different ESD stressing. While all being oxide rupture type, the defect locations vary, i.e., Figure 6.8a for defect at internal NMOS device under HBM tests, Figure 6.8b for damage in internal PMOS device by MM zapping, and Figure 6.8c for defect in internal NMOS device from CDM stressing [13]. Figure 6.9 shows that gate oxide defects occur either in NMOS ESD protection device under MM pulsing (a) or in an internal NMOS device by CDM ESD zapping (b), while contact spiking damage (c) appears in ESD protection resistor under HBM stressing for the same audio chip in 1.5µm CMOS process [13]. These two examples clearly demonstrate that ESD failure mechanisms may vary for the same IC part depending upon ESD zapping modes. Figure 6.10 shows a light emission ESD defect image for a two-finger ggNMOS ESD protection structure in a

(a) (b) (c)

Figure 6.8 Similar gate oxide ESD defects are observed at various data bus locations in a communication chip under different ESD stressing modes: (a) in NMOS for HBM, (b) in PMOS for MM, and (c) in NMOS for CDM [13] (*Reproduced here by kind permission of the ESD Association and authors*).

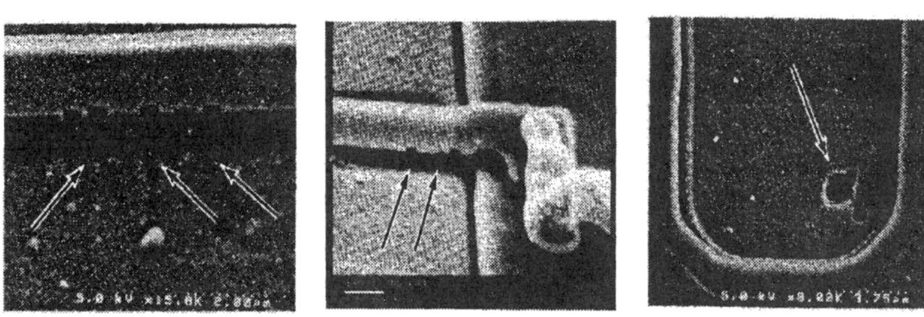

(a) (b) (c)

Figure 6.9 Different ESD zapping tests resulted in different ESD failure signatures in an audio chip, i.e., (a) oxide defect in a NMOS ESD protection device for MM, (b) oxide damage in an internal NMOS for CDM, and (c) contact spiking damage in an ESD protection resistor for HBM [13] (*Reproduced here by kind permission of the ESD Association and authors*).

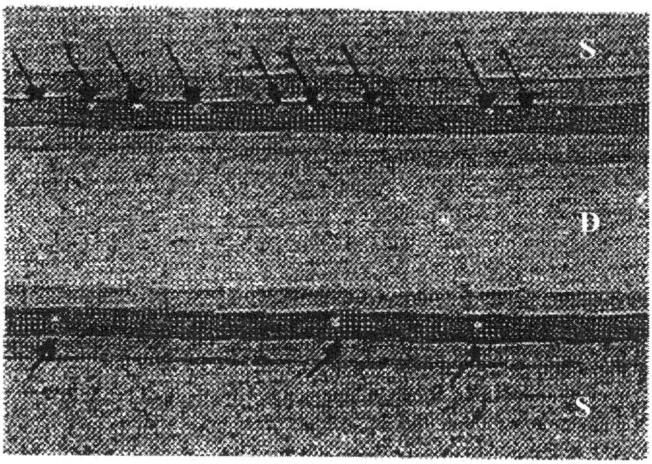

Figure 6.10 Fairly evenly-distributed hot spots appear in the drain contact to gate regions in both fingers of the two-finger ggNMOS ESD protection structure with LDD-blocking, indicating uniform turn-on, an improvement over non-uniform triggering in its with-LDD counterpart version [14] (*Reproduced here by kind permission of the ESD Association and authors*).

0.35μm CMOS technology. Fairly evenly distributed defects (hot spots marked by arrows) were observed between the drain contacts and gates at both fingers. This uniform triggering feature is attributed to blocking of LDD (lightly-doped drain) implant, a commonly used technique to boost ESD protection by reducing electric field density at drain junction corners. In comparison, hot spots were observed in one finger only for the same circuit without using LDD-blocking technique [14].

Another classic ESD failure signature is ESD damages in metal interconnect due to joule heating, occurring in both aluminium (Al) and copper (Cu) interconnect technologies. Figure 6.11 shows Al extrusion type ESD defects in Ti/Al/Ti interconnects in a 0.25μm CMOS technology, where dielectric cracking occurs due to over-heating in Al metal and melt Al fills the dielectric cracks [15]. Various metal damages in a cladded-Cu interconnect caused by HBM stressing, e.g., dielectric cracking, extrusion, blistering and displacement, are given in Figure 6.12 in a 0.18μm CMOS case study [15]. It is clear that Cu interconnect is more ESD-robust than Al

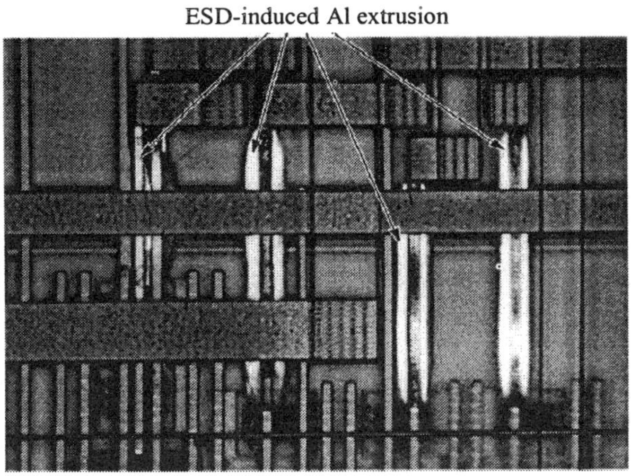

Figure 6.11 ESD-induced metal extrusion defect in Al interconnects [15] (*Reproduced here by kind permission of IEEE and authors*).

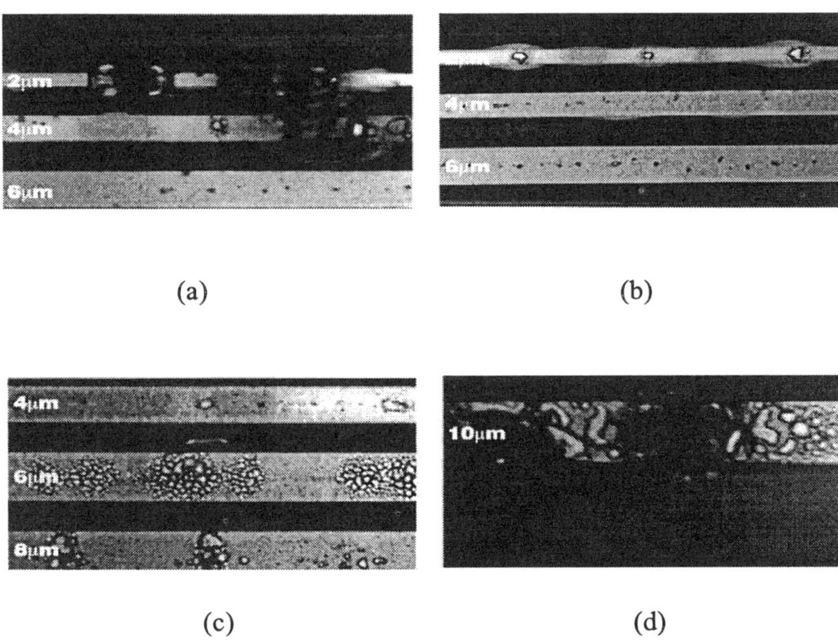

(a) (b)

(c) (d)

Figure 6.12 ESD-induced defects in Cu interconnects [15] (*Reproduced here by kind permission of IEEE and authors*).

interconnect, a highly valuable feature that will result in less ESD-metal-induced parasitic capacitance in a circuit if design properly, which will be discussed in Chapter 8 for ESD-circuit interactions. Technology scaling has its influences in ESD damages in metal interconnects as well. In one study [16], two 5μm metal power supply lines with original thickness of 0.5μm and scaled thickness of 0.45μm used in N=/n-well diode ESD protection structures were investigated. Under HBM ESD stressing, the thicker metal did not show any damage up to 10kV, while ESD defects in the forms of metal vaporization and electro-thermal migration were observed in the scaled metal structure, as shown in Figure 6.13. Metal interconnect design for ESD protection is fairly tricky. Cares should be excised in selecting metal widths both to ensure adequate ESD robustness and to reduce ESD-metal-induced parasitic capacitances simultaneously.

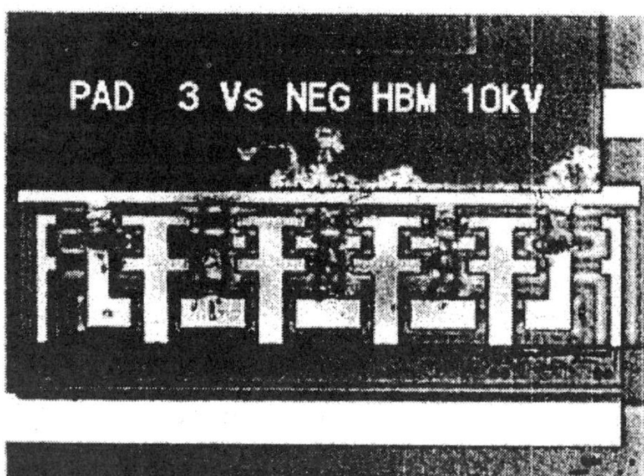

Figure 6.13 HBM ESD-induced metal vaporization and electro-thermal migration defects in the scaled metals in a diode ESD protection structure [16] (*Reproduced here by kind permission of the ESD Association and authors*).

The following two FA examples illustrate how FA analysis helps to understand the ESD protection structure triggering procedures and the ESD failure mechanisms. In the first case [17], light emission microscopy technique was employed to investigate the triggering procedures of ggNMOS ESD protection structures in a 0.5μm silicided LDD CMOS technology. Both silicide-blocking and LDD-blocking techniques were used to improve ESD protection performance. Two ggNMOS structures with and

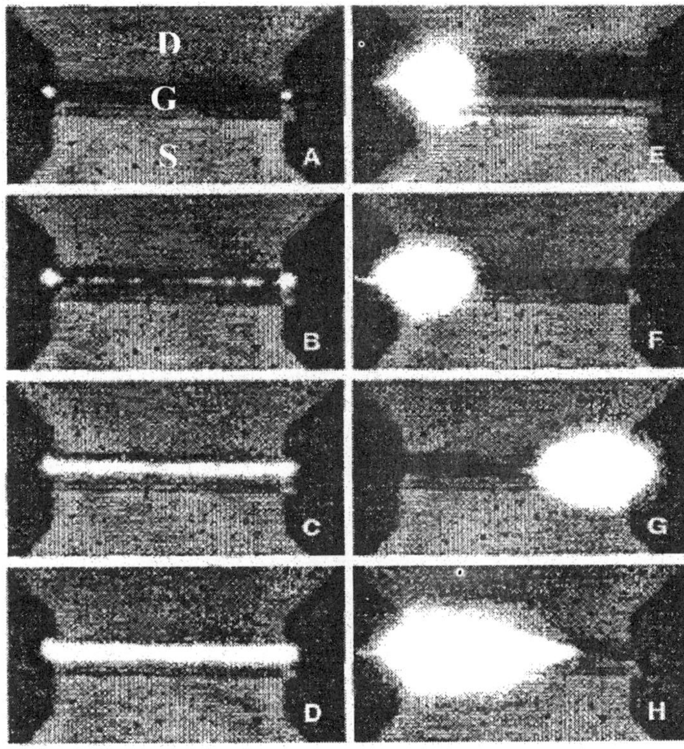

(a)

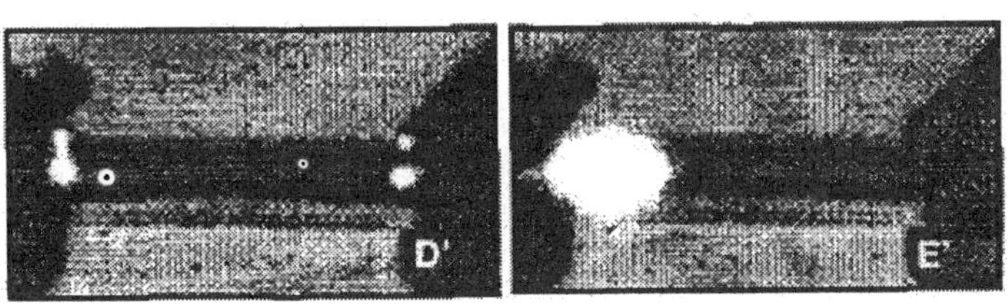

(b)

Figure 6.14 EMMI light emission images under DC stressing for a non-LDD ggNMOS (a) and a LDD ggNMOS (b) ESD protection devices. The non-LDD ggNMOS features uniform triggering reflected by a continuous light emission line in image D, while non-uniform turn-on is observed for the LDD device as shown in image D' [17] (*Reproduced here by kind permission of the ESD Association and authors*).

without LDDs were studied for their triggering uniformity by applying both DC and TLP stressing. The EMMI light emission images are given for the DC-stressed triggering sequences for non-LDD (a) and LDD (b) ggNMOS, respectively. For the non-LDD ggNMOS device in Figure 6.14a, light spots first appear at the corners in image A at the bias of 10.2V and 20μA. Continuous increase in the bias leads to the light emission region developed from light spots into to continuous light lines, i.e., image B (450μA) to C to D (10.9V and 2mA). At stage D, high enough an avalanche current is available to trigger the ggNMOS and drives it into the snapback region. After snapback, very bright and cornered light emission mega spots appear at the ends of the channel and light spot hopping between the two ends were observed as avalanche current continuously increase, corresponding to image E to H (60mA). Therefore, the trigger voltage of this ggNMOS is about $V_{t1} \sim$ 10.9V. A continuous light emission line before triggering indicates a uniform turn-on mechanism for this non-LDD ggNMOS. Initiation of light spots, i.e., avalanche currents, at the corners is due to the strong electric field density formed at the sharp junction corners. The hopping of light spots after snapback is attributed to the thermal nature causing the triggered regions switching. For the LDD ggNMOS structure as shown in Figure 6.14b, light emission also initiated at the junction corners. However, the light spots (avalanche current) remain in the corner regions all the way to the triggering threshold (image D') and into the snapback region. Similarly, strong and confined light emission (image E') was observed at either end in the snapback region. It is clear that the LDD ggNMOS device features non-uniform triggering as oppose to the uniform turn-on in the non-LDD case. This difference in triggering sequences is due to doping profile variation at the drain junction corner in the channel. This observation backs the commonly accepted argument that LDD doping degrades ESD protection performance and an LDD-blocking mask should be used in sub-micron ESD protection design in order to improve ESD robustness.

In the second FA analysis example, an NMOS structure, implemented in a 0.5μm salicided LDD CMOS technology, was examined for its evolution from soft to hard ESD failures under TLP and HBM stressing [18]. Figure 6.15 shows the TLP-stressed ESD damages by optical microscope and AFM technique. A soft defect initiated at the drain end under a 200ns TLP pulse is shown in Figure 6.15a. After a stronger 500ns TLP stress, the defect developed into a full D-S Si filament hard failure as shown in Figure 6.15b, which is also confirmed by the AFM images in Figure 6.15 c and d.

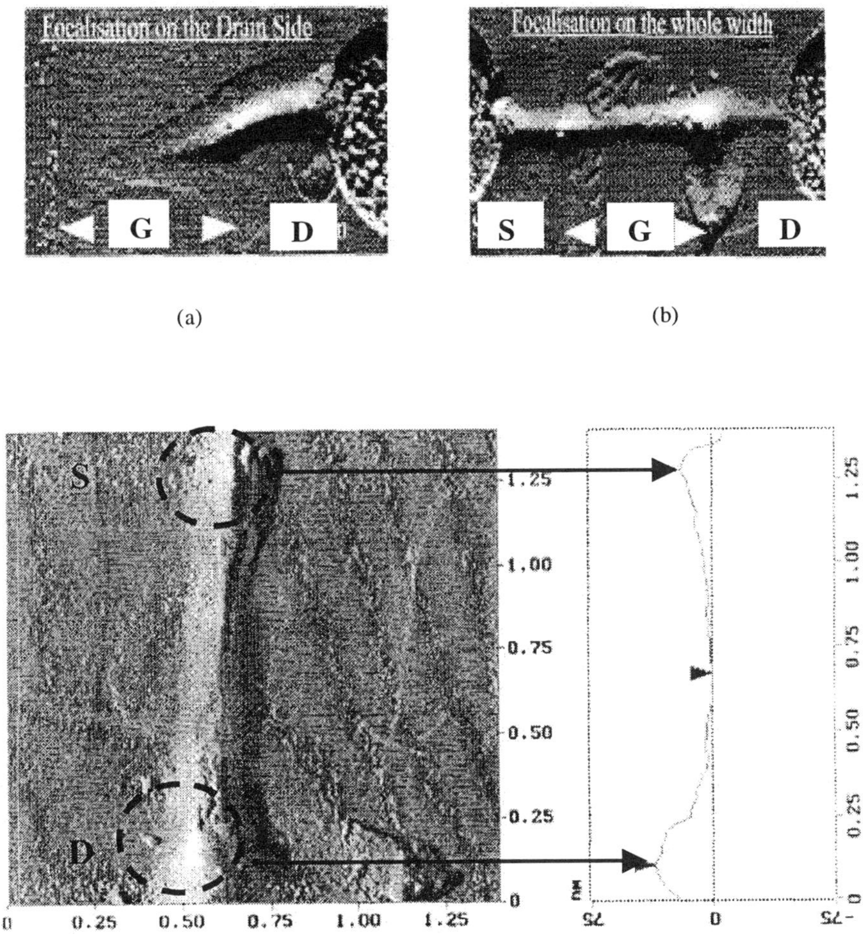

Figure 6.15 Optical microscopy and AFM images show the soft to hard failure evolution of a MOS device under TLP stresses, (a) a soft defect initiates at drain end under a 200ns TLP pulse, (b) a 500ns TLP pulse develops the soft defect into a hard failure featuring a D-S filament, which is confirmed by AFM image (c) and AFM section analysis (d) [18] (*Reproduced here by kind permission of the ESD Association and authors*).

Atomic force microscopy, or AFM, technique can be very helpful in detecting extremely tiny ESD defects. Figure 6.16 shows AFM images for a CDM ESD-induced latent defect in a CMOS gate circuit [19]. A tiny defect hole (~0.19μm x 0.14μm) in oxide sidewall is located in the AFM image. The corn-shaped defect hole is clearly presented in the high-magnification AFM image of Figure 6.16b. In another study of HBM ESD failure of P+/n-well diode ESD protection structures for a DRAM chip in a shallow trench isolated (STI) CMOS technology [20], a subsurface defect underneath the STI dielectric is located in an SEM image after de-processing as shown in Figure 6.17a. This subsurface ESD defect can be viewed more clearly in the AFM image given in Figure 6.17b. More ESD FA analysis application examples are discussed in the following sections.

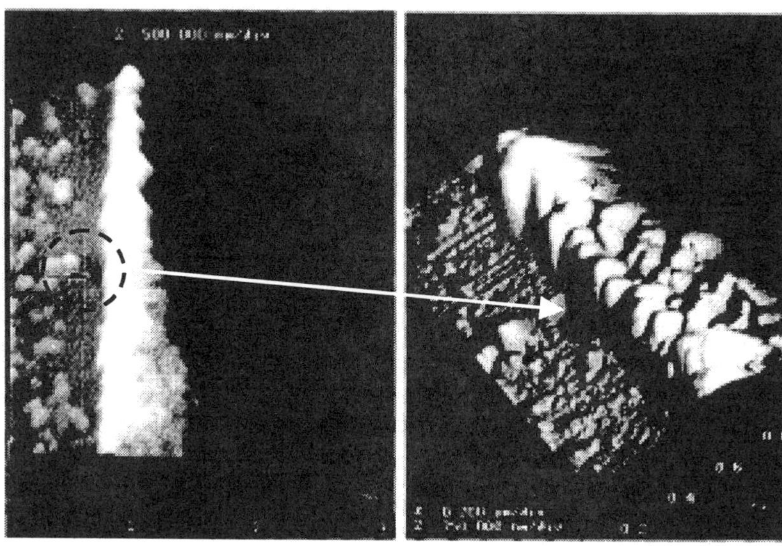

(a) (b)

Figure 6.16 AFM images of a CDM ESD-induced latent defect in oxide sidewall in a CMOS input gate circuit. The high-magnification image is shown in (b) for the defect hole in (a) [19] (*Reproduced here by kind permission of the ESD Association and authors*).

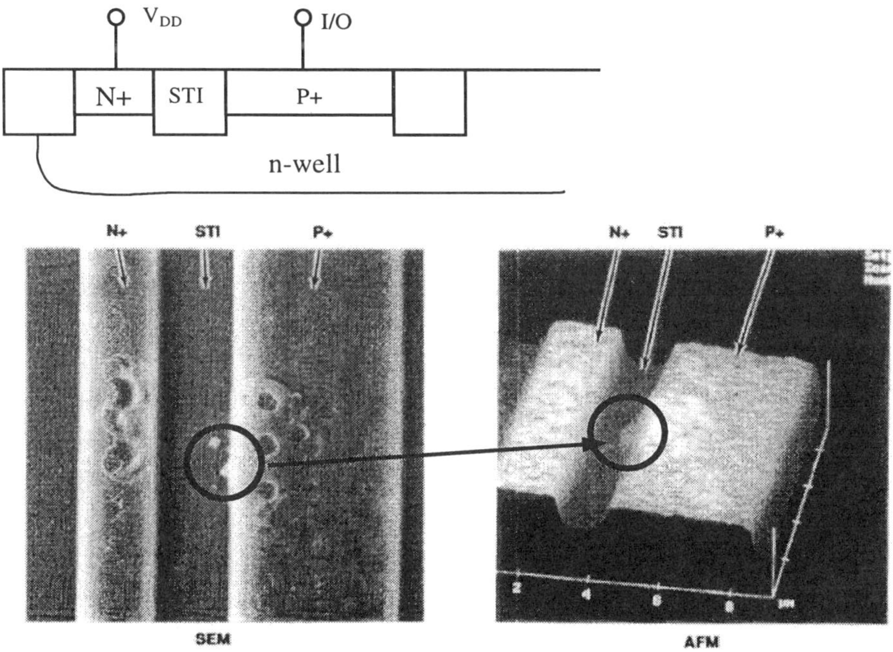

(a) (b)

Figure 6.17 An after-deprocessing SEM image shows a subsurface ESD defect in a P+/n-well ESD protection diode for a DRAM chip (a), with its high-resolution image given by an AFM image shown in (b) [20] (*Reproduced here by kind permission of the ESD Association and authors*).

6.4. ESD FA CORRELATION

As discussed in Chapter 2, there are a variety of different ESD test models proposed and used for ESD stressing tests, e.g., HBM, MM, CDM, TLP and IEC, etc. Accordingly, many types of ESD zapping testing systems, homemade or commercial, are used in the field. It is fairly natural for one to try to compare the ESD protection levels, often given in the ESD failure voltage threshold level denoted by ESDV in kilo-voltage (kV), of IC parts produced by different IC vendors for their ESD robustness. However, such comparison usually leads to complete confusions. For example, using different ESD test models will produce very different ESDV kV numbers for the same IC parts. It would be nice to establish a valid correlation between these different ESD test models, preferably using a formula by IC designers. Unfortunately, such valid correlation is still something on the wish list currently. One main barrier in this matter is associated with the un-reproducibility of ESD test results due to equipment parasitic effects, particularly in the MM and CDM cases.

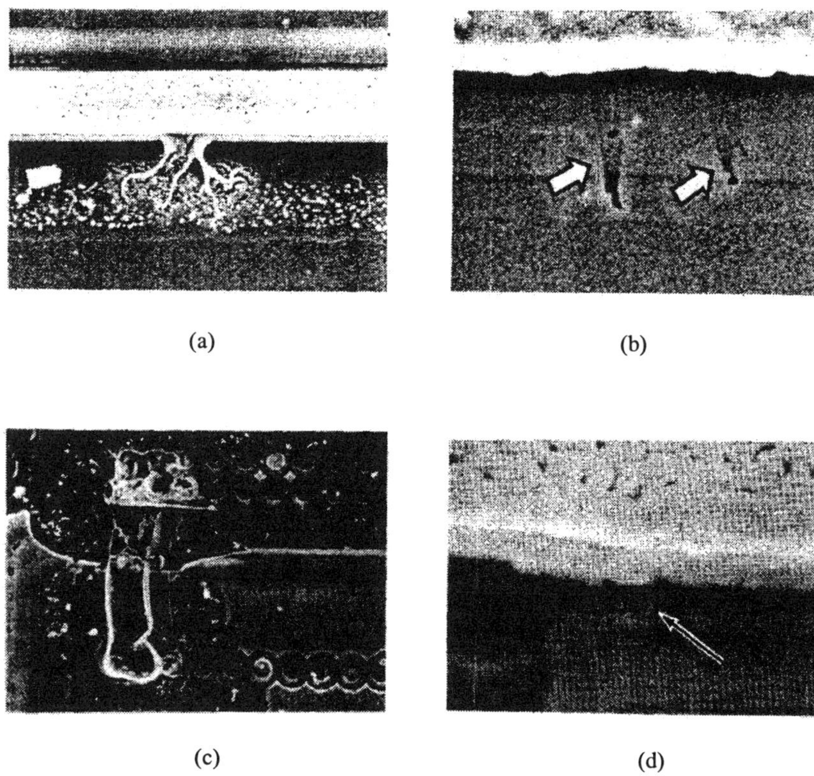

Figure 6.18 FA images for Device 1 under different ESD stresses, (a) Poly-Si extrusion in NMOS ESD protection device from HBM tester 1, (b) oxide defect in the same internal NMOS by HBM tester 2, (c) metal melting in NMOS ESD protection by MM stressing, and (d) gate oxide defect and Poly-Si filaments in internal PMOS by CDM zapping [13] (*Reproduced here by kind permission of the ESD Association and authors*).

In the first case, a set of different IC chips was evaluated using HBM, MM and CDM zapping testers at different user locations [13]. The IC parts evaluated include ASIC disk driver chips, audio chips, data communication interface ICs, and automotive control chips, made in 0.9μm, 1.2μm and 1.5μm CMOS technologies. The ESD damage images for Device 1, an automotive control IC in 1.2μm CMOS, are given in Figure 6.18. It is found that different HBM ESD-induced failures occurred using two different HBM zapping testers, i.e., poly-silicon extrusion in an NMOS ESD protection device by HBM tester 1, as illustrated in Figure 6.18a, and gate oxide damage in an internal NMOS device by HBM tester 2, as shown in Figure 6.18b. MM ESD zapping leads to metal melting damage in NMOS ESD protection as shown in Figure 6.18c, while Figure 6.18d shows both oxide defect and poly-silicon filaments in an internal PMOS under CDM stresses. FA analysis of Device 2 shows similar oxide defects under all HBM, MM and CDM stresses, however, at different locations on the chip. The FA images were discussed previously in Figure 6.8, where image (a) showing oxide defect in an internal NMOS under HBM zapping, image (b) illustrating oxide damage in an internal PMOS by MM stressing, and image (c) depicting a defect in NMOS from CDM test. Evaluation of Device 3 show different ESD failures due to different ESD zaps. The corresponding FA signatures were given in Figure 6.9, where MM stress leads to damage in the protection resistor and oxide failure in NMOS ESD protection structure as in image (a), CDM zapping results in oxide defect in an internal NMOS as in image (b), and HBM stressing leads to contact spiking damage in the ESD protection resistor as in image (c). The above results indicate that a simple correlation may not exist between HBM, MM and CDM failure signatures. Hence, ESD failure signature is a poor correlation indicator. Fortunately, further analysis suggests that, if using ESD protection level, i.e., the ESDV number in kV, as the indicator, a fairly good correlation factor of 0.92 is obtained between the HBM and MM data. However, there is still no correlation when CDM test is involved.

In a second FA example, the HBM ~ TLP correlation was investigated for a group of NMOS devices in 0.5μm and 0.35μm CMOS technologies [21]. A good matching between the ESD and TLP stressing data was observed for the 0.5μm NMOS devices, which is in agreement with the FA analysis where the same contact spiking damages were found in both cases as shown by its SEM image in Figure 6.19. However, poor correlation was obtained between the HBM and TLP results for the 0.35μm NMOS devices. The FA analysis found different failure signatures for HBM and TLP stresses, i.e., poly-Si filament and slight Si melting in drain region due to HBM zapping as shown in Figure 6.20a, and a D-S Si filament damage after

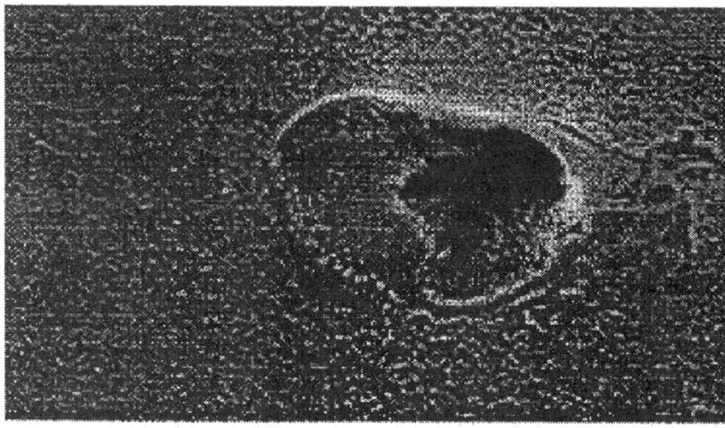

Figure 6.19 The same contact spiking defects occur in the NMOS devices in 0.5μm and 0.35μm processes under HBM and TLP stresses, indicating a good HBM-TLP correlation [21] (*Reproduced here by kind permission of the ESD Association and authors*).

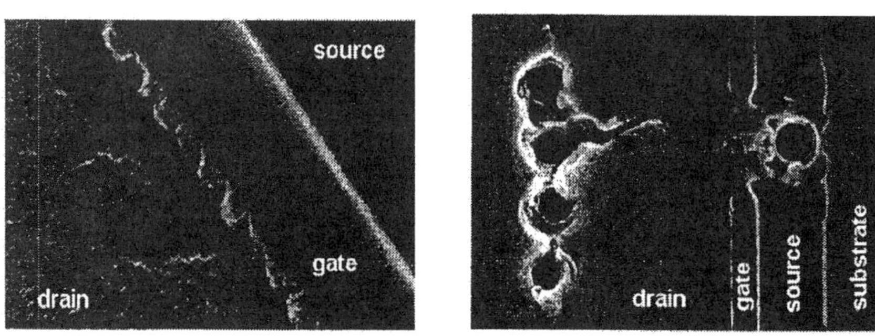

(a) (b)

Figure 6.20 Different ESD failure signatures are observed in a 0.35μm NMOS under 1.3A HBM (a) and 1.5A TLP stresses (b), resulting in a poor HBM-TLP correlation [21] (*Reproduced here by kind permission of the ESD Association and authors*).

TLP stress illustrated in Figure 6.20b. This study touches one of the most arguable topics in ESD protection field – adopting TLP as an ESD evaluation standard. As discussed in Chapter 2, TLP emerges as a very useful ESD stressing technique that provides the highly desirable instantaneous I-V characteristics during stressing tests that help one to design ESD protection circuits. It would be wonderful if TLP can be developed as a standard ESD test model, provided a valid correlation being found between HBM and TLP measurements. Some studies do suggest that good correlation exist between HBM and TLP tests, provided cares being excised in the measurements [22-25]. For example, one study performed for ggNMOS ESD protection structures in 0.25μm CMOS technology at both wafer and package levels concluded that using the simplified formula given by Equation 6.1 leads to a poor correlation between the HBM and TLP data. However, a fairly good HBM-TLP correlation can be obtained if a series resistance in the discharging channel, R_S, is included in the formula as given by Equation 6.2 [25].

$$V_{t2,HBM} = I_{t2,TLP} R_{HBM} \tag{6.1}$$

$$V_{t2,HBM} = I_{t2,TLP} (R_S + R_{HBM}) \tag{6.2}$$

where R_{HBM} = 1.5kΩ is the human body resistance used in the HBM model, $I_{t2,\ TLP}$ is the second breakdown current obtained by TLP measurements, and $V_{t2,\ HBM}$ is the equivalent HBM ESD protection level. The new factor R_S can be obtained by either the least square fitting [26] or the lumped element model method [27]. With so many uncertainties in publications, one ought to be cautious in dealing with the data correlating work. A good rule-of-thumb is always conducting some first-hand statistical data analysis for confirmation.

6.5. LATENT ESD FAILURE

Hard ESD damages are usually what an IC designer concerns about during ESD protection circuit design, which is already a complex subject. Latent ESD damage, though often overlooked, is yet another real ESD reliability problem. Unfortunately, things seem to be in total chaos in dealing with latent ESD failures. The biggest issue is the lack of a solid quantitative analysis method in dealing with latent ESD failures. Usually, statistical approaches are used in characterising soft ESD damages. A latent ESD failure can be defined as a time-dependent ESD damage that results in noticeable, however still acceptable, deterioration in critical circuit specifications. Consequently, reduction in lifetime may be expected. Typically observed electrical parameters associated with latent ESD damages include junction leakage currents, gate oxide leakage current, and threshold voltage drift due to low-level ESD stresses. These electrical parameters have been quantitatively analysed using time-dependent dielectric breakdown, or, TDDB, lifetime technique [28, 29] or oxide noise signature analysis [19], among many other often-controversial studies [30-32]. In a latent ESD failure analysis study for a CMOS technology with a gate oxide of 190Å, two types of latent defects were examined by applying DC and HBM stressing to NMOS devices, i.e., latent gate oxide defects due to trapped charges that lead to increase in gate leakage current and drift of threshold voltage, and soft damages in the drain junction that cause increase in drain leakage current [28]. The study was performed by directly stressing the gate and drain by DC and HBM methods. Similar latent failures were observed in both cases. These soft defects in gate oxide and drain junction, which result in leakage current increases, were evaluated in terms of TDDB lifetimes. The latent damage induced leakage currents range from a few pA to several hundreds of mA. The study concludes that no noticeable lifetime reduction occurs due to gate oxide latent defects until gate leakage reaching to up to several hundreds of nA, as shown in Figure 6.21. Thermal annealing at a temperature of 200°C can significantly recover the gate leakage current degradation, but has no effect on the lifetime. On the other hand, strong lifetime reduction was observed associated with the latent defects in the drain junction even at a lower current stress level as indicated in Figure 6.21 as well. Thermal annealing has no effect on both drain leakage and lifetime in the drain damage case. In another study of latent oxide damage for ultra thin oxide films of 2.2-4.7nm by DC and TLP stressing, the soft oxide defects are believed to be associated with TDDB electron trap generation and charge trapping in oxide films [29]. The latent oxide damage is then correlated to the stress-induced leakage current, denoted as SILC, which is defined as the difference between the after-stress leakage current and before-stress leakage current, i.e., SILC $\equiv I_{AS} - I_{BS}$. The trap generation rate is believed to be proportional to the SILC/ I_{BS} [33]. It is found that the SILC/ I_{BS} is strongly stress-pulse-width dependent as shown in Figure 6.22, indicating that DC stressing data can not be used to predict latent ESD oxide failures. A simple formula is then proposed to describe the latent ESD damages in the oxide,

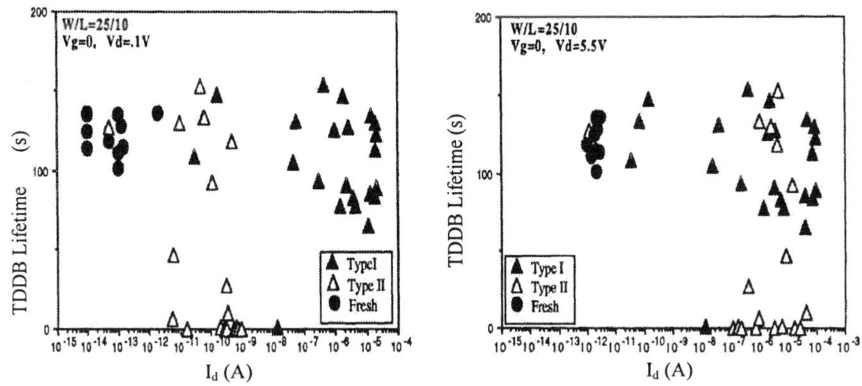

(a) (b)

Figure 6.21 Measured TDDB lifetime ~ I_d curves for NMOS devices in fresh and stressed conditions indicate no noticeable lifetime reduction due to latent oxide defects (type I), however, significant lifetime degradation associated with soft drain damage (type II). (a) measured at V_d = 0.1V, and (b) tested at V_d=5.5V [28] (*Reproduced here by kind permission of the ESD Association and authors*).

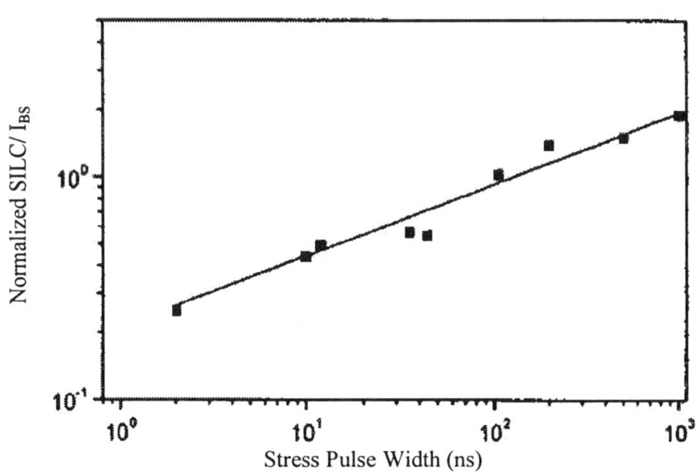

Figure 6.22 A normalized SILC/I_{BS} ~ t curve measured for a stressed thin oxide shows dependence of latent ESD damage induced gate leakage on stress pulse width [29] (*Reproduced here by kind permission of the ESD Association and authors*).

$$SILC / I_{BS} \propto t^{n} \tag{6.3}$$

where t is the stress pulse width and the fitting factor n is extracted as $n \approx 0.33$.

6.6. ESD FAILURE MODELING AND CRITERIA

Failure analysis is certainly a very useful and informative technique in ESD protection design. However, as readers may have been asking many times while reading, FA is too qualitative an approach that offers no parametric numbers to guide ESD protection circuit design. Indeed, IC design is a quantitative task that talks in numbers. It is extremely important to extract mathematical models from experimental observations that can be used to thoroughly understand ESD protection operation physics and to direct predictive ESD protection circuit design. Design predication is the ultimate goal in ESD protection research.

ESD modelling has two aspects: analytical ESD failure models that accurately describe ESD protection operation and ESD failure mechanisms, and ESD protection device models that enable predictive ESD protection design simulation. Accurate ESD protection device modelling is particularly difficult because of the high current and thermal effects involved in ESD operation as compared to regular SPICE device modelling. Research in high-current ESD device modelling is still in its infancy currently. Some proposed ESD protection device models would be discussed in Chapter 8 that covers ESD protection design simulation methodology. The following discussion deals with mathematical ESD modelling accounting for ESD failure procedures.

ESD failures can be modelled by either thermal model or electrothermal model. Thermal modelling is relatively simple, where device electrical parameters are assumed to be temperature-independent. The power consumption of an ESD protection device under ESD transients converts into heat generation, which leads to lattice temperature increase in Si. ESD failure threshold is usually defined as a critical temperature in concerned materials, i.e., a melting temperature of 1412°C for Silicon, a eutectic temperature of 550°C in aluminium, or, a melting temperature of 1084°C in copper interconnects. A typical analytical ESD failure model follows.

Assume a known heat source as a parallelepiped in an ESD protection device, say, an NMOS, with dimensions of a, b, and c as shown in Figure 6.23. The heat conduction equation is then solved under proper boundary conditions for device internal temperature distribution in time domain for a given power consumption [34]. Since the thermal diffusion length is no longer than a few μm for silicon in the ESD pulse time scale (<150ns), which is much smaller than the wafer/die dimensions, the heat equation can be solved for an infinite medium. Of course, this condition is not valid in the up-vertical direction because IC devices are very close to the Si surface that is covered by very poor thermal-conductive dielectrics. Hence, a reasonable boundary condition set should be $-\infty<x<\infty$, $-\infty<y<\infty$, $-\infty<z\leq 0$. An approximate solution to the heat equation is then given by,

$$P_f = \left(\frac{A}{t_f} + \frac{B}{\sqrt{t_f}} + \frac{C}{\lg t} + D\right)(T_C - T_o) \qquad (6.4)$$

where T_c is the critical temperature corresponding to the ESD failure threshold, T_o is the ambient temperature, P_f is defined as the power-to-failure

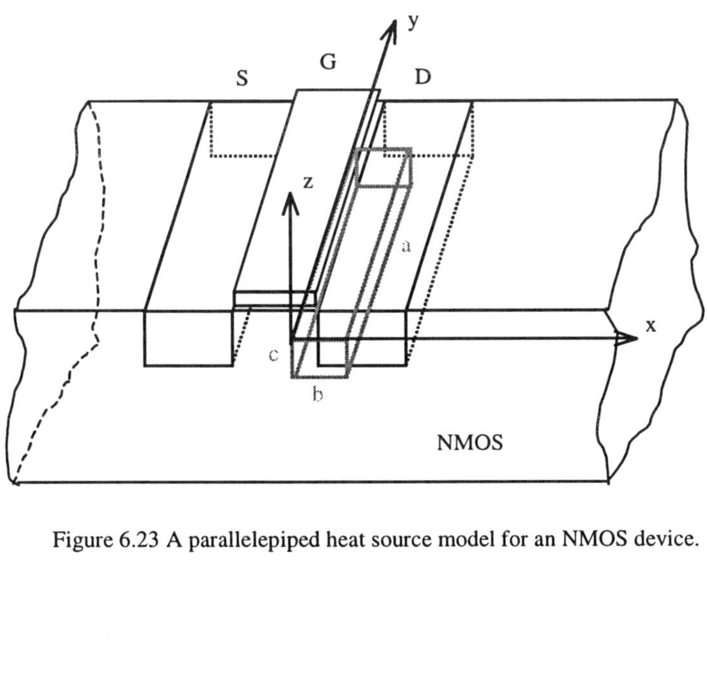

Figure 6.23 A parallelepiped heat source model for an NMOS device.

for the power required to cause ESD failure under an ESD transient with a fixed duration, t_f is defined as the time-to-failure for the time needed to cause ESD failure under a specific ESD injection power, and A, B and C are related coefficients. The thermal diffusion time constants associated with the three dimensions are defined as,

$$t_a = \frac{a^2}{4\pi D},\ t_b = \frac{b^2}{4\pi D},\ t_c = \frac{c^2}{4\pi D}, \tag{6.5}$$

where D is the thermal diffusion constant of silicon. A commonly seen four-segment ESD failure model is then depicted in Figure 6.24. Should readers prefer more rigorous solution, the heat equation can be solved using the Green's function method to give the following solutions:

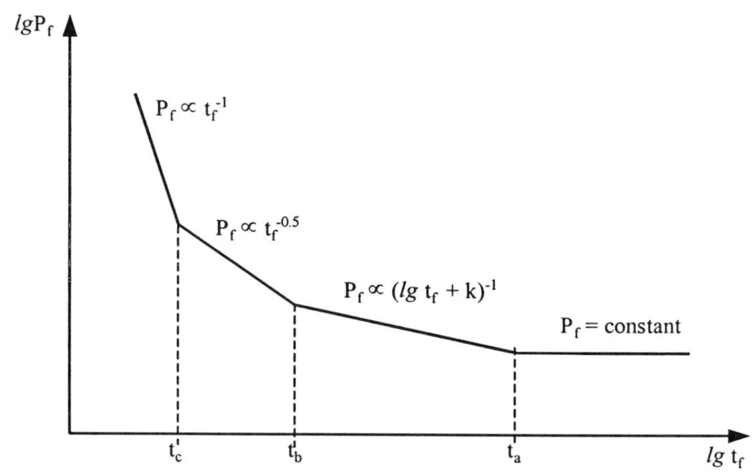

Figure 6.24 A four-segment ESD failure model.

$$P_f = \frac{A' abc(T_c - T_o)}{t_f}, \qquad 0 < t_f < t_c, \quad (6.6)$$

$$P_f = \frac{B' ab(T_c - T_o)}{\sqrt{t_f} - \sqrt{t_c}/2}, \qquad t_c < t_f < t_b, \quad (6.7)$$

$$P_f = \frac{C' a(T_c - T_o)}{\lg(t_f / t_b) - 2 - c/b}, \qquad t_b < t_f < t_a, \quad (6.8)$$

$$P_f = \frac{D' a(T_c - T_o)}{\lg(a/b) + 2 - c/2b - \sqrt{t_a / t_f}}, \qquad t_a < t_f, \quad (6.9)$$

where A', B', C' and D' are related coefficients. This nice $P_f \sim t_f$ model derived from thermal modelling may be good for after-math study of ESD failure behaviours; unfortunately, it has little values for predicting ESD protection circuit design.

For more accurate ESD failure modelling, electrothermal models should be used where electrical parameters are treated as temperature-dependent and thermal equation and semiconductor device physics equations are solved in a coupled approach. The second breakdown, or, thermal runaway, threshold derived from the electrothermal equations can be used as the onset of ESD failure. While being more accurate, this electrothermal modelling is certainly much more complex and time-consuming.

6.7. SUMMARY

In this chapter, ESD failure analysis techniques are discussed. FA analysis is important in the sense that FA results help circuit designers to better understand ESD protection failure mechanisms and to avoid making similar design mistakes repeatedly. Typical ESD failure signatures, such as, silicon filament, metal interconnect burnout, contact spiking, gate oxide rupture, etc, are discussed. Case examples are given to show how FA analysis techniques can be used in practical design debugging. Latent ESD failure phenomena are described using real examples. Analytical ESD device failure modelling is presented as well. While FA analysis is basically a statistic technique, understanding typical ESD failure mechanisms would certainly increase the success rate of practical ESD protection circuit design.

REFERENCES

[1] Amerasekera, A. and Duvvury, C., *ESD in Silicon Integrated Circuits*, New York, Wiley, 1995.

[2] Richards, B. and Footner, P., *The Role of Microscopy in Semiconductor Failure Analysis, Oxford Microscopy Handbook, Vol. 5,* Oxford University Press, 1992.

[3] Rasras, M., De Wolf, I. Groeseneken, G. Maes, H., Vanhaeverbeke, S. and De Pauw, P., "A Simple, Cost Effective, and Very Sensitive Alternative for Photon Emission Spectroscopy", *Proc. ISTFA Symp.*, 1997, pp.153-157.

[4] Bude, J., Sano, N. and Yoshii, A., "Hot-Carrier Luminescence in Si", *Phusical Review B, Vol. 11,* 1992, pp.5848-5856.

[5] Deboy, G. and Kolzer, J., "Fundamentals of Light Emission from Silicon Devices", *Semicon. Science Tech., No. 9,* 1993, pp.1017-1032.

[6] Kolzer, J., Boit, C. Dallmann, A., Deboy, G., Otto, J. and Weinmann, D., "Quantitative Emission Microscopy", *J. Appl. Phys., 71(11),* 1992, pp.R23-R41.

[7] Russ, C. Bock, K., Rasras, De Wolf, I., Groeseneken, G. and Maes, H., "Non-Uniform Triggering of gg-nMOSt Investigated by Combined Emission Microscopy and Transmission Line Pulsing", *Proc. EOS/ESD Symp.,* 1998, pp.177-186.

[8] Clark, N., Parat, K., Maloney, T. and Kim, Y., "Melt Filaments in n+pn+ Lateral Bipolar ESD Protection Devices", *Proc. EOS/ESD Symp.,* 1995, pp.295-303.

[9] Smith, J., "A Substrate Triggered Lateral Bipolar Circuit for High Voltage Tolerant ESD Protection Applications", *Proc. EOS/ESD Symp.*, 1998, pp.63-71.

[10] Amerasekera, A., van den Abeelen, W., van Roozendaal, L., Hannemann, M. and Schofield, P. "ESD Failure Modes: Characteristics, Mechanisms, and Process Influences", *IEEE Trans. Elec. Dev., Vol. 39, No. 2,* February 1992, pp.430-436.

[11] Duvvury, C. and Amerasekera, A., "Advanced CMOS Protection Device Trigger Mechanisms During CDM", *Proc. EOS/ESD Symp.,* 1995, pp.162-174.

[12] Kiefer, S. Milburn, R. and Rackley, K., "EOS Induced Polysilicon Migration in VLSI Gate Array", *Proc. EOS/ESD Symp.,* 1993, pp.123-127.

[13] Kelly, M., Diep, T., Twerefour, S., Servais, G., Lin, D. and Shah, G., "A Comparison of Electrostatic Discharge Models and Failure Signatures for CMOS Integrated Circuit Devices", *Proc. EOS/ESD Symp.,* 1995, pp.175-185.

[14] Richier, C., Maene, N., Mabboux, G. and Bellens, R., "Study of the ESD Behaviour of Different Clamp Configurations in a 0.35um CMOS Technology", *proc. EOS/ESD Symp.*, 1997, pp.240-245.

[15] Voldman, S., "The Impact of Technology Scaling on ESD Robustness of Aluminium and Copper Interconnects in Advanced Semiconductor Technologies", *IEEE Trans. Components, Packaging, Manufacturing Tech. – Part C, Vol. 21, No 4*, October 1998, pp.265-277.

[16] Voldman, S. and Gross, V., "Scaling, Optimisation and Design Considerations of Electrostatic Discharge Protection Circuits in CMOS Technology", *Proc. EOS/ESD Symp.*, 1993, pp251-260.

[17] Russ, C., Bock, K., Rasras, M., de Wolf, I., Groeseneken, G. and Maes, H., "Non-Uniform Triggering of ggnMOS Investigated by Combined Emission Microscopy and Transmission Line Pulsing", *Proc. EOS/ESD Symp.,* 1998, pp.177-186.

[18] Salome, P., Leroux, C., Mariolle, D., Lafond, D., Chante, J., Crevel, P. and Reimbold, G., "An Attempt to Explain Thermally Induced Soft Failures During Low Level ESD Stresses: Study of the Differences Between Soft and Hard NMOS Failures", *Proc. EOS/ESD Symp.,* 1997, pp.337-345.

[19] Colvin, J., "The Identification and Analysis of Latent ESD Damage on CMOS Input Gates", *Proc. EOS/ESD Symp.*, 1993, pp.109-116.

[20] Never, J. and Voldman, S., "Failure Analysis of Shallow Trench Isolation ESD Structures", *Proc. EOS/ESD Symp.*, 1995, pp.273-288.

[21] Stadler, W., Guggenmous, X., Egger, P., Gieser, H. and Mussshoff, C., "Does the ESD Failure Current Obtained by Transmission-Line Pulsing Always Correlate to Human Body Model Tests?" *Proc. EOS/ESD Symp.,* 1997, pp.366-372.

[22] Notermans, G., de Jong, P. and Kuper, F., "Pitfalls When Correlating TLP, HBM and MM Testing", *Proc. EOS/ESD Symp.,* 1998, pp.170-176.

[23] Barth, J., Verhaege, K., Henry, L., and Richner, J., "TLP Calibration, Correlation, Standards, and New Techniques", *Proc. EOS/ESD Symp.,* 2000, pp.85-96.

[24] Barth, J. and Richner, J., "Correlation Considerations: Real HBM to TLP and HBM Testers", *Proc. EOS/ESD Symp.,* 2001, pp.453-460.

[25] Lee, M., Liu, C., Lin, C., Chou, J., Tang, H., Chang, Y. and Fu, K., "Comparison and Correlation of ESD HBM Obtained Between TLPG, Wafer-Level, and Package-Level Tests", *Proc. EOS/ESD Symp.,* 2000, pp.105-110.

[26] Giordano, A. and Hsu, F., Chapter 2, *Least Square Estimation with Applications to Digital Signal Processing*, Wiley, New York, 1985.

[27] Verhaeg, K., Roussel, P., Groeseneken, G., Maes, H., Gieser, H., Russ, C., Egger, P., Guggenmos, X. and Kuper, F., "Analysis of HBM ESD Testers and Specifications Using a 4th Order Lumped Element Model", *Proc. EOS/ESD Symp.,* 1993, pp.129-137.

[28] Song, M., Eng, D. and MacWilliams, K., "Quantifying ESD/EOS Latent Damage and Integrated Circuit Leakage Currents", *Proc. EOS/ESD Symp.,* 1995, pp.304-310.

[29] Wu, J., Juliano, P. and Rosenbaum, E., "Breakdown and Latent Damage of Ultra-Thin Gate Oxides under ESD Stress Conditions", *Proc. EOS/ESD Symp.,* 2000, pp.287-295.

[30] Greason, W., Kucerovsky, Z. and Chum, W., "Latent Effects due to ESD in CMOS Integrated Circuits: Review and Experiments", *IEEE Trans. Industry Appls., Vol. 29,* 1993, pp.88.

[31] Woodhouse, J. and Lowe, K., "ESD Latency: A Failure Analysis Investigation", *Proc. EOS/ESD Symp.,* 1988, pp.47.

[32] Tunnicliffe, M., Dwyer, V. and Campbell, D., "Latent Damage and Parametric Drift in Electrostatically Damaged MOS Transistors", *J. Electrostatics,* 1993, pp.91.

[33] DiMaria, D., "Electron Energy Dependence of Metal-Oxide-Semiconductor Degradation", *Appl. Phys. Lett., Vol. 75, No. 16,* 1999, pp.2287-2295.

[34] Dwyer, V., Franklin, A. and Campbell, D., "Thermal Failure in Semiconductor Devices", *Solid State Electronics, Vol. 33,* May 1990, pp.553-560.

Chapter 7

LAYOUT AND TECHNOLOGY INFLUENCES ON ESD PROTECTION CIRCUIT DESIGN

On-chip ESD protection circuitry design is an extremely complex design task that involves multiple level coupling effects, i.e., process-device-circuit-electro-thermal couplings. In addition to selecting proper ESD discharging mechanisms, both process technology and parasitic effect play major roles in the final ESD protection performance [1-3]. The parasitic-effect dependence of ESD protection operation suggests strong geometry sensitivity in ESD protection circuit performance. Hence, IC layout plays a vital role in success of ESD protection circuit design. A large percentage of ESD protection design failures may actually come from layout issues. On the other hand, as IC technologies advance, featuring continuous dimension shrinkage and new process techniques, such as, LDD (lightly-doped drain), silicidation, SiGe layer, and SOI (silicon-on-insulator), etc, both positive and negative impacts are expected on ESD protection operation. Consequently, ESD protection circuit design is, unfortunately, non-portable in nature, as oppose to the common misunderstanding in current IC design practices. This chapter discusses typical influences of circuit layout and process technology factors on ESD protection circuit design.

7.1. LAYOUT *vs.* ESD PROTECTION

Parasitic effect is one of the most troublesome, however, common, issues in ESD protection circuit design. As stated previously, the principle of ESD protection is to set up a safe discharging channel between any two pads of concerns. It is usually not too difficult for one to design an ESD protection structure working this way. Moreover, such an ESD protection structure is often optimised to survive very strong ESD transients, well, by itself. However, in real world design, it is not unusual at all for one to experience early ESD zapping failures when conducting chip level measurements. One main root cause to such kind of pre-mature ESD failure is associated with the turn-on of parasitic devices either within the designed ESD protection networks or inside the core circuit being protected. It is certainly true that nothing would be wrong for a parasitic discharging element to shunt an ESD pulse in parallel with the designed ESD protection structures. However, if its discharging impedance is much lower than that in the designed ESD discharge channel, a parasitic device may be easily destroyed during ESD events, leaving the designed ESD protection network being merely a silicon consumer. The fact that many ESD protection structures actually rely on parasitic device operation, e.g., parasitic lateral NPN BJT in a ggNMOS ESD protection structure, makes the situation even worse, since one ought to utilize some useful parasitic devices while suppress all unwanted parasitic effects. Unfortunately, IC designers do not enjoy any significant margin room between the desired and undesired parasitic operations to play along with in practical design. These parasitic devices are often due to inappropriate layout design, e.g., spacing, etc. On the other hand, thermal damages to designed ESD protection structures usually come from current crowding induced local over-heating stemming from layout discontinuity, for example, at the corners and edges of diffusion layers. Therefore, layout optimisation is extremely important in practical ESD protection circuit design. It would be wonderful if one can rely on software tools in performing ESD design layout checking. Unfortunately, a comprehensive and reliable ESD layout checking CAD tool does not exist at the time of this writing. Eye-checking still plays a vital role in this matter. This section briefly discusses typical ESD layout techniques.

7.2. REGULAR LAYOUT FOR ESD PROTECTION

Diodes, bipolar transistors, MOEFET transistors and SCRs are the most widely used ESD protection structures. This section discusses basic layout considerations for these building block elements.

Diode ESD protection structures can be implemented in many different fashions. Any existing junction diode may serve as an ESD protection device, typically being N^+/P^+, p-well/n-well, N^+/p-well, P^+/n-well, etc. However, their discharging efficiencies vary substantially. ESD protection diodes can be either vertical or lateral ones, which are drawn as diffusion bottom diode or sidewall diode in layout. The main concerns are certainly discharging impedance and heat dissipation. A vertical N^+/p-well ESD protection diode layout is illustrated in Figure 7.1, where the N^+ diffusion

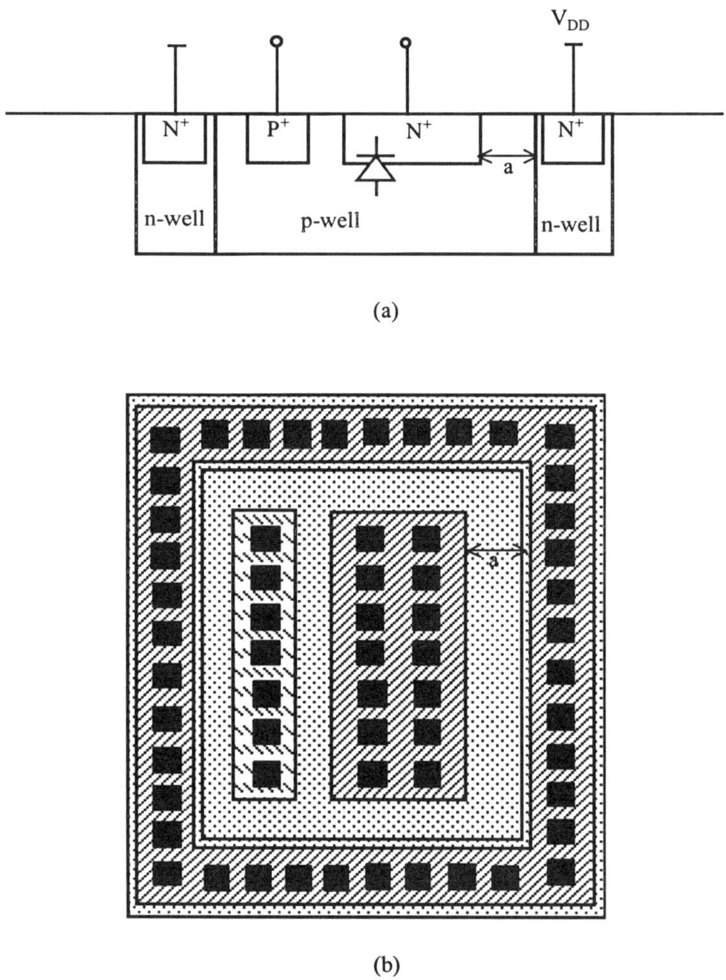

Figure 7.1 A layout for a vertical p-well/ N^+ ESD protection diode where the discharging area is the N^+ diffusion bottom. Special consideration in setting up the spacing is needed to avoid high discharging impedance and lateral NPN parasitic BJT with n-well guard-ring.

bottom defines the current discharging area that is selected based upon the desired ESD protection level, because the N^+ sidewall diode is much smaller. The lateral spacing between the N^+ diffusion and P^+ pick-up is a critical factor affecting the discharging resistance. As in most cases, a guard-ring, sometimes, dual guard-rings, should be used. Taking into consideration of the parasitic lateral NPN BJT, an adequate spacing between the N^+ diffusion and the n-well guard-ring, marked as *a* in the illustration, is required. As always, layout uniformity is important for ESD protection performance. For example, all contacts must be placed evenly in the diffusions to avoid possible current crowding effect. In addition, an enough number of contacts are needed according to the peak ESD transients. Junction sidewall diode is another commonly used ESD protection diode structure. Since heat dissipation can only occurs in the downward direction, a deep junction is highly desired. Hence, a p-well/n-well diode serves as a very efficient sidewall ESD protection diode. Because only the peripheral of a sidewall diode contributes to ESD transient conduction, while the diffusion bottom only increase undesired parasitic junction capacitance, an interdigitated layout design is preferred as shown in Figure 7.2 for a p-well/n-well diode. The critical concerns in layout design of ESD protection diodes are heat dissipation capacity, discharging impedance, effective discharging area and parasitic devices. In selecting junction diffusion layer, doping profile and physical size, one ought to consider all the ESD-induced parasitic parameters as well, such as parasitic capacitance and noise, in order to balance the needs for adequate ESD protection and minimum ESD-associated influences on circuits.

As discussed in Chapter 3, many ESD protection structures are based upon BJT transistor operation for snapback I-V characteristics. It is evident that an original BJT transistor performs better as an ESD protection device compared to any parasitic BJT devices because of its high current gain. BJT ESD protection structures can be designed as either vertical devices or lateral ones. A vertical BJT transistor in bipolar or BiCMOS process technologies is a preferred ESD discharging device because of its superior bipolar gain and low discharging impedance. A lateral BJT ESD protection structure is often used in advanced CMOS technologies, where no dedicated base diffusion is available for making vertical bipolar transistors. Such a lateral BJT ESD protection structure is inherently much less efficient in discharging ESD transients because of its relative wider base width limited by photolithography resolutions, hence lower bipolar gain. A typical layout for a vertical NPN ESD protection structure is given in Figure 7.3. Its lateral base extension resistance has big influence on current gain. Proper layout arrangement, such as a type of CBEBC diffusion pattern, helps to reduce the

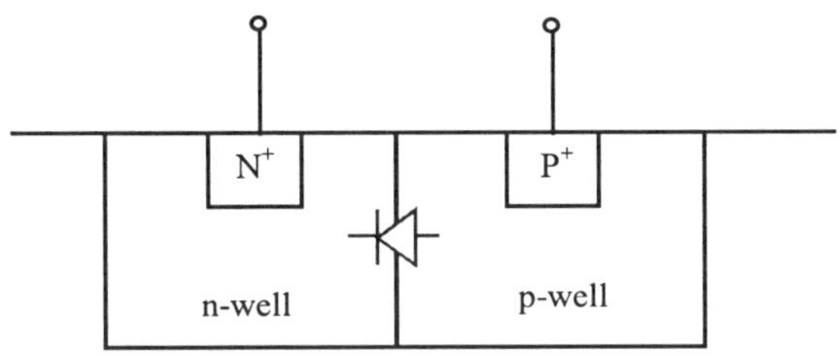

(a)

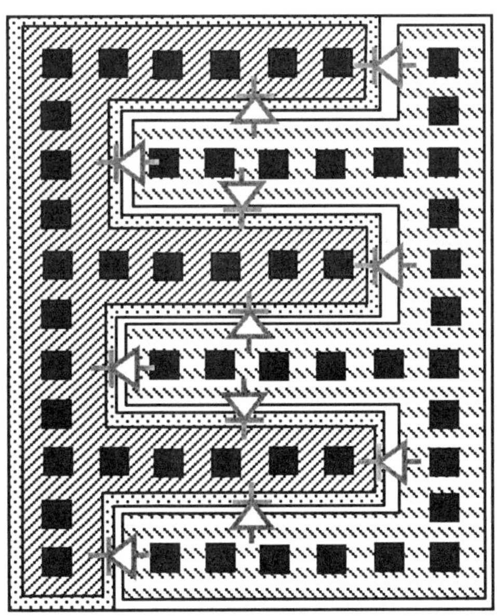

(b)

Figure 7.2 A layout for a p-well/n-well sidewall ESD protection diode uses an interdigitated format to maximize the effective ESD discharging area.

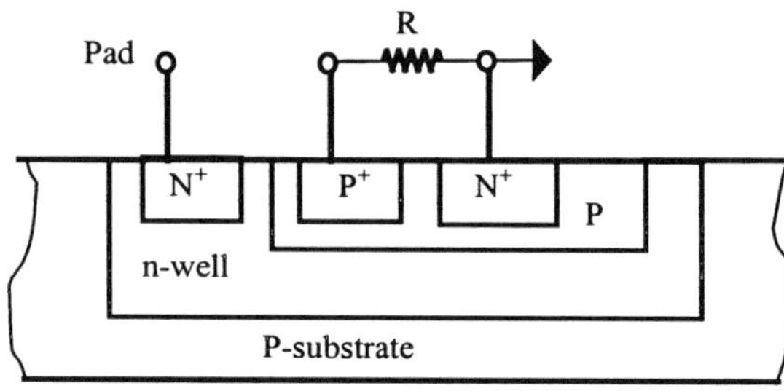

(a)

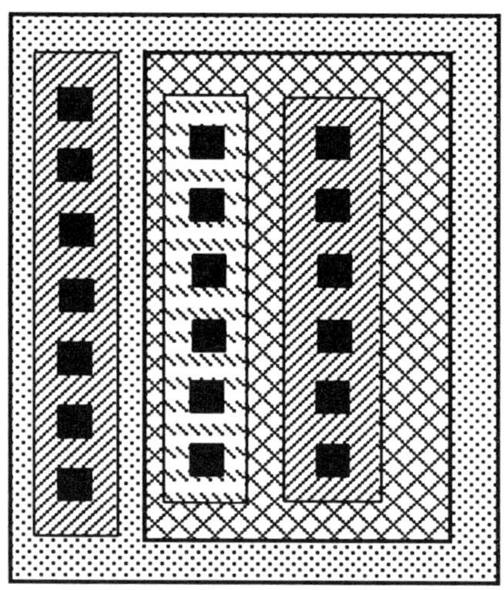

(b)

Figure 7.3 A typical layout for a vertical NPN BJT ESD protection structure.

terminal extension resistances. Uniform current flow is a main factor to concern about in diffusion finger, contact and metal layout. Spacing between guard-ring diffusion and internal diffusions must be adequate to avoid parasitic lateral BJT turn on.

MOSFET ESD protection structures are widely used in CMOS technologies. A lot of layout tricks are reported for MOS ESD protection structures [1-3]. Figure 7.4 illustrates a typical layout of a ggNMOS ESD

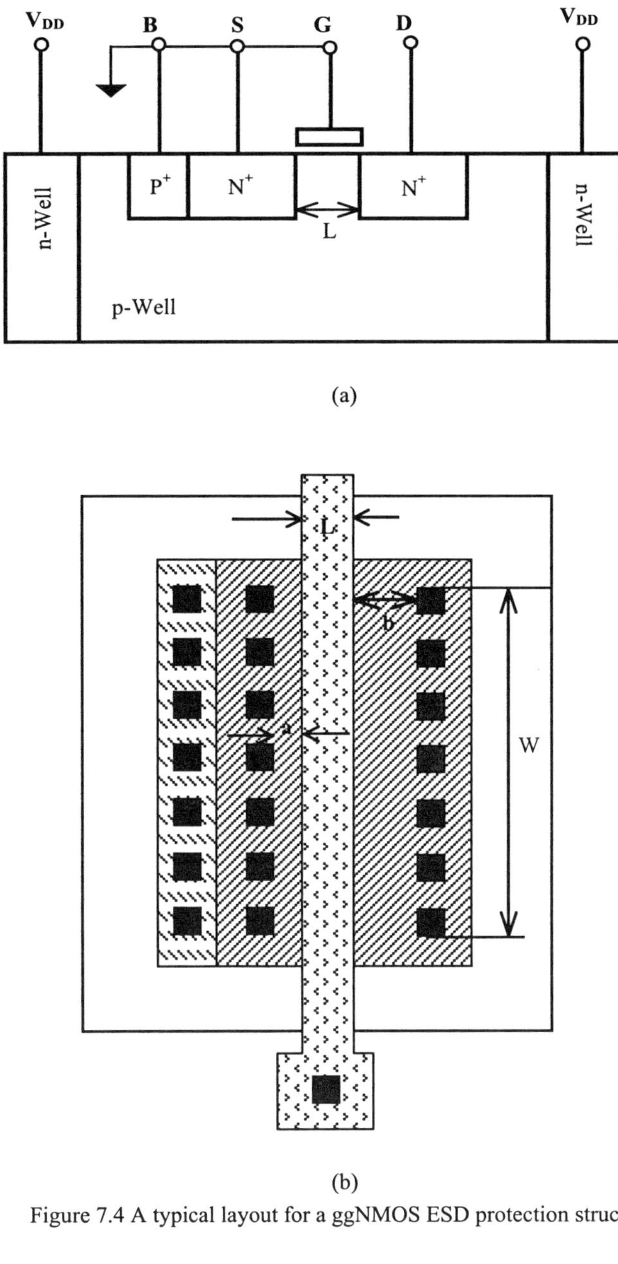

(b)

Figure 7.4 A typical layout for a ggNMOS ESD protection structure.

protection structure, where the channel length, L, equivalent to the base width of the parasitic NPN in ESD protection operation, is a critical dimension that should be minimized, provided no punch-through breakdown occurs. The ggNMOS device width, i.e., W, is determined according to the desired ESD protection level. Depending upon the process technologies used, MOSFET layout can be very different. For example, a large drain-contact-to-gate spacing, *b*, can boost ESD protection in non-silicide technology, but not in modern silicide CMOS. While, the source-contact-to-gate spacing, *a*, should be minimized to lower the discharging impedance. These special layout issues will be discussed in next section. In addition, some common rules apply in metal interconnect routing and contact/via placement to achieve uniform current flow, which will also be discussed in the following section. Figure 7.5 shows a layout for a thick-gate NMOS ESD protection, typically used in a primary-secondary ESD protection scheme, where deep n-well source and drain diffusions are used for better heat dissipation. Being able to use deep S/D diffusions is a main benefit for using a thick-gate MOS ESD protection device. The n-well to n-well spacing, L, should be minimized for better NPN current gain. However, this effective base width, limited by field oxide encroachment, is fairly wide, making the parasitic lateral NPN transistor less efficient in discharging ESD currents. This limitation may be overcome in advanced CMOS using shallow trench isolation (STI) technique.

SCR-based ESD protection structures are very area-efficient and can achieve very high ESD protection level. With many effective anti-latch-up techniques available, an SCR ESD protection structure shall be a preferred option for ESD protection in advanced technologies, because it consumes much smaller silicon asset and produces far less ESD-induced parasitic effects. Figure 7.6 shows a classic SCR ESD protection structure layout that can be readily implemented in a CMOS technology. In layout design, critical numbers include the p-well enclosure of cathode n^+ diffusion, *a*, which may ignite unwanted punch-through under positive ESD pulse, and the p-well to anode p^+ diffusion, *b*, which is the effective lateral PNP base width. The sum of *a* and *b* also affect ESD discharging impedance and SCR current gain product, therefore influence ESD protection performance significantly. In addition, since higher ESD protection is expected, the metal interconnect routing and contact/via placement become far more a concern in SCR ESD protection structure layout than in other cases. A straight current flow as shown in Figure 7.6 is preferred. However, since its triggering is ignited by relative high n-substrate to p-well junction avalanche breakdown, a classic SCR ESD protection structure does not find wide applications due to its high V_{t1}. This high triggering-V problem can be resolved by either inserting an n^+

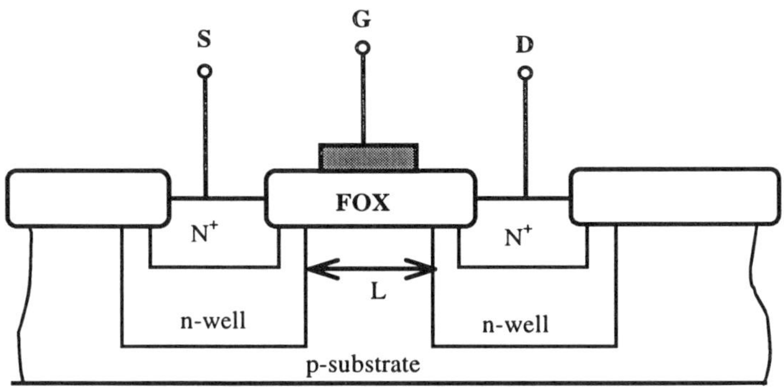

(a)

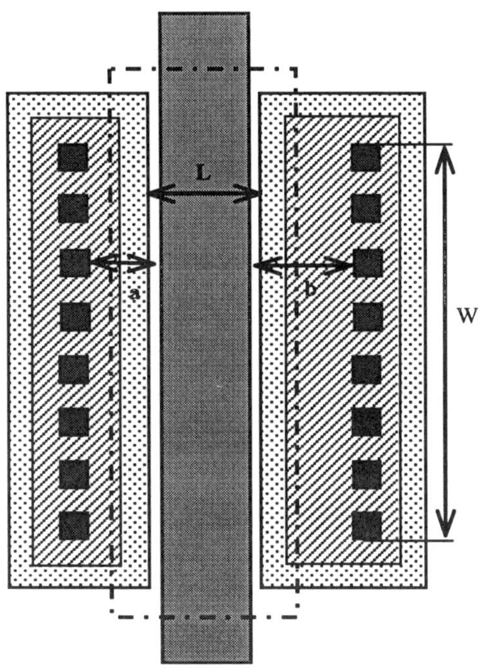

(b)

Figure 7.5 A typical layout for a thick-gate NMOS ESD protection structure.

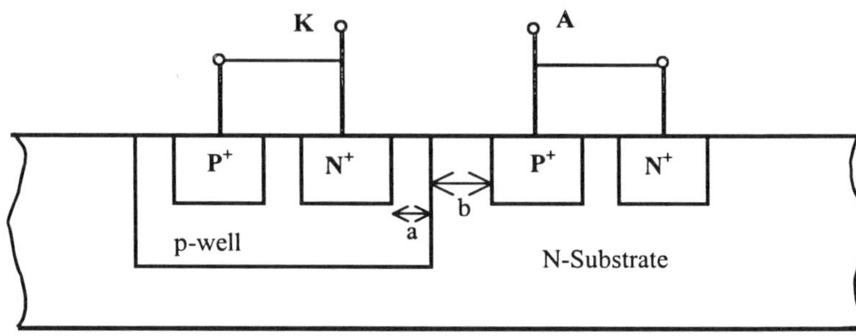

(a)

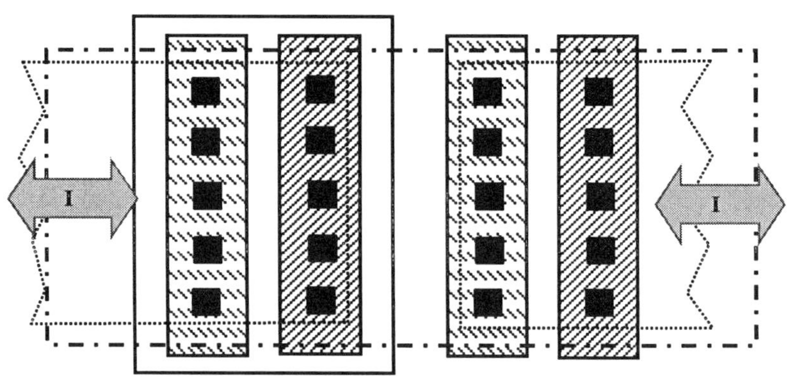

(b)

Figure 7.6 A typical layout for a classic SCR ESD protection structure in CMOS prefers straight current flow along metal lines.

diffusion across the n-substrate to p-well boundary or using a ggNMOS device to trigger the SCR device, which was discussed in Chapter 4. The latter two low-triggering versions of SCR ESD protection structures are depicted by Figure 7.7 and Figure 7.8, respectively.

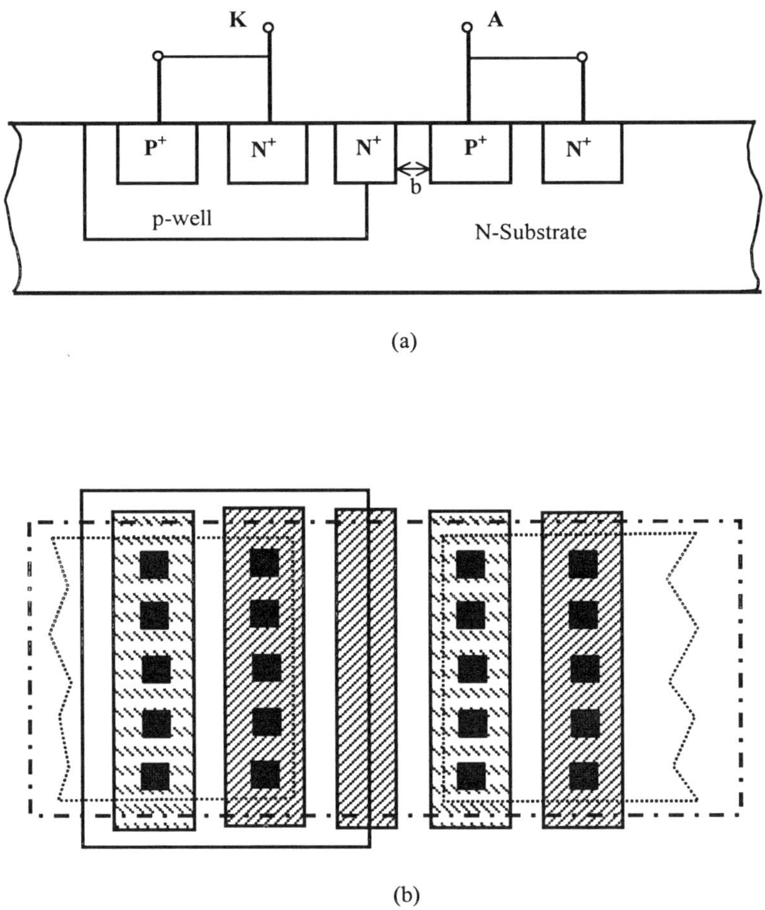

Figure 7.7 A layout for a low-triggering SCR ESD protection structure introduced in Chapter 4 uses an n^+ insertion to reduce the avalanche breakdown voltage of the SCR device, therefore lower the triggering voltage.

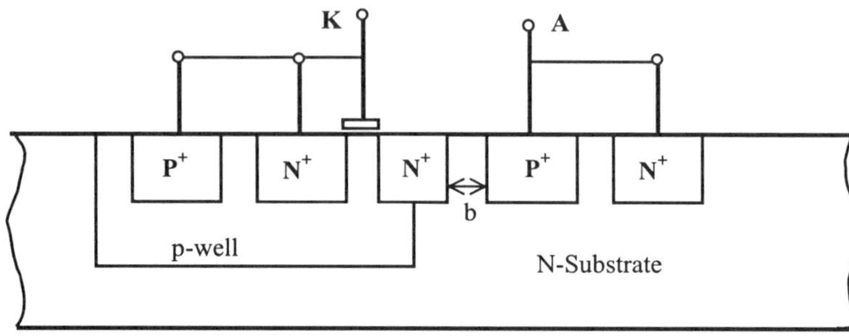

(a)

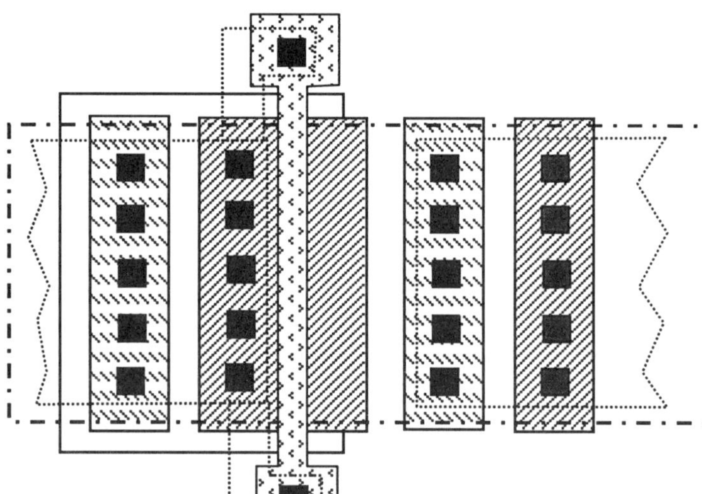

(b)

Figure 7.8 A typical layout for a ggNMOS-triggered low-triggering SCR ESD protection structure discussed in Chapter 4.

7.3. SPECIAL LAYOUT FOR ESD PROTECTION

The typical layout schemes introduced in the previous section may ensure operational ESD protection structures to some extent, however, a lot more subtle layout issues must be taken care of in order to realize an optimal design and to achieve high ESD protection performance. This is the same as conducting analog and mixed-signal circuit design, where designing a working circuit is really not a big deal, however, making a good design that is optimised for speed, power dissipation, size, and accuracy, etc, all together is a totally different story. Such knowledge-based artistic design skills distinguish a top IC designer from a so-so circuit design employee. In ESD protection structure layout design, critical factors that must be considered include triggering mechanisms, holding threshold, discharging impedance, current flow uniformity and heat dissipation profiles, etc. A well-thought layout usually serves to boost ESD protection performance substantially in these regards.

Take the widely used MOSFET ESD protection structure, shown in Figure 7.4, as an example. A well-proven layout guideline suggest that, in order to achieve the best ESD protection, the source-contact-to-gate spacing (SCGS) should be minimized according to process design rules, while the drain-contact-to-gate spacing (DCGS) should be around 5μm [4-9]. This SCGS/DCGS layout design rule for NMOS ESD protection structures apply to non-silicide CMOS technologies. The rationale behind came from research data that found that ESD failure voltage threshold (ESDV) increases monotonically with the DCGS until somehow reaching to a saturation point at around a DCGS of 4~6μm, while no such improvement was observed when varying the SCGS spacing. Several factors may play roles in this magic DCGS~ESDV formula. First, it is believed that heat generation usually occurs at the lower corner of a drain junction facing the channel, which may spread into the drain contact regions and cause thermal damages in contacts and metals. Hence, an adequate DCGS spacing effectively removes the hot spot away from the contact and metal. Second, the series resistance of the drain extension serves as a ballasting resistor that ensure uniform current distribution and, therefore, uniform triggering in multiple-finger NMOS ESD protection structures. The uniform turn-on across NMOS fingers leads to maximum ESDV value to be achieved for the given structure. On the negative side, excessive ballasting resistance in drain extension increases discharging impedance and heat generation, which results in lower ESD robustness of a given ESD protection structure. Increase of drain series resistance also tends to degrade ESD protection in

CDM mode. Therefore, a balance in selecting DCGS spacing is imaginable, which translates into an optimal DCGS value. On the other hand, no ESD protection improvement is observed by increasing the SCGS spacing. This observation seems to be reasonable because no contact and metal melting is expected at the source end since the heating source is located far away at the drain end. Increasing the SCGS spacing will, however, increase the total resistance in discharging channel. Figure 7.9 shows a modified layout for an

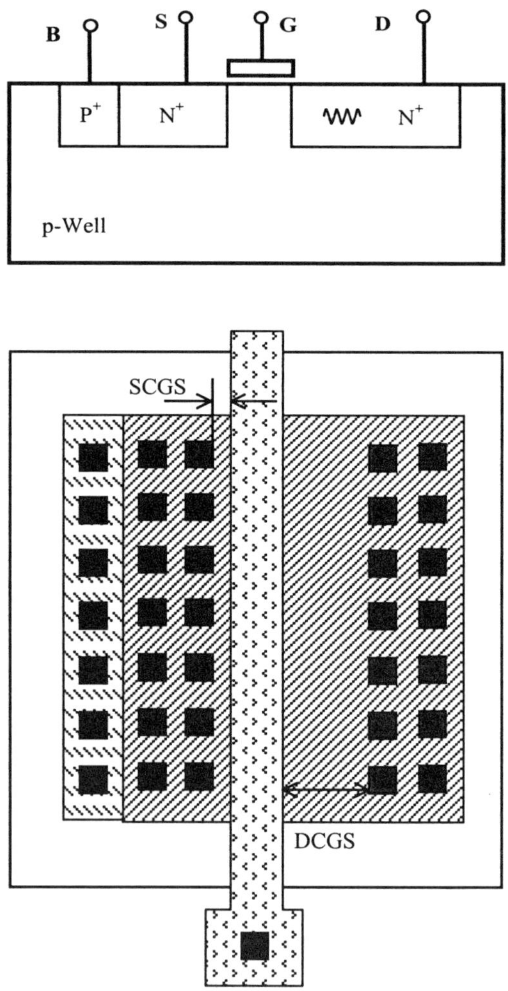

Figure 7.9 A layout for an NMOS ESD protection shows large DCGS and small SCGS designed to boost the ESD robustness.

NMOS ESD protection structure that accounts for the DCGS and SCGS factors. Improvement of ESD robustness by increasing DCGS spacing becomes less significant in CMOS technologies using LDD technique and the benefit vanishes almost completely in advanced silicided CMOS technologies. This is due to the fact that the ballasting effect disappears as silicidation reduces the drain extension resistance dramatically. It is worth to point out that this minimum-SCGS/large-DCGS layout rule comes from experimental data for those above-0.25μm CMOS technologies. It is necessary to evaluate the detail ESDV~DCGS/SCGS effect in practical design by use of ESD simulation [4]. In very-deep-sub-micron IC technologies, the layout rules may change due to effects, such as, ballistic phonon scattering induced micro-overheating phenomenon, etc. For example, initial study suggests that the minimum-SCGS rule may no longer hold in sub-0.18μm CMOS technologies because the source contact may be too close to the heat source at the drain end. Once again, one is reminded of

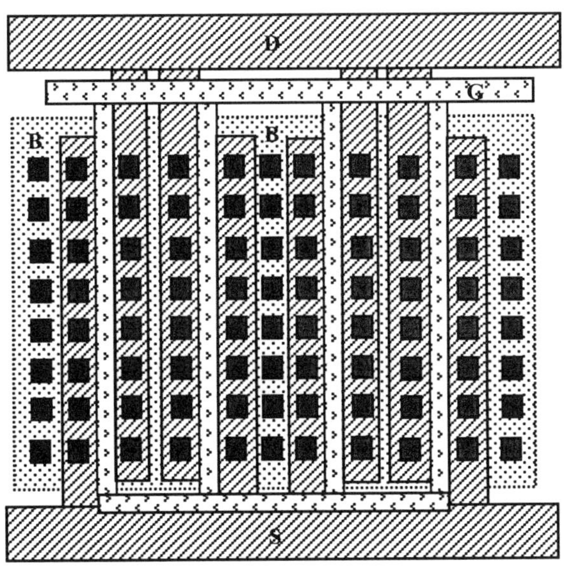

Figure 7.10 A BSGD-DGSBSGD-DGSB layout pattern for a multiple-finger NMOS ESD protection structure achieves the best ESD protection result due to uniform current distribution [10] (*Reproduced here by kind permission of Elsevier Science and authors*).

not to over-trust existing ESD protection design rules. Design customisation and ESD design simulation are essential in any successful design of ESD protection circuits.

Using multiple-finger structure to achieve high ESD protection performance assumes a linear ESDV to device size (or, finger number) relationship. This assumption may hold under a uniform turn-on condition, for which a large DCGS spacing and sufficient ballasting resistance are in place. In addition, other layout consideration is necessary to ensure uniform current distribution. For example, proper layout patterns for the source (S), drain (D), gate (G) and body (B) fingers help to distribute ESD currents evenly across fingers. Study shows that a BSGD-DGSBSGD-DGSB finger layout pattern achieves better ESD protection level compared to other layout

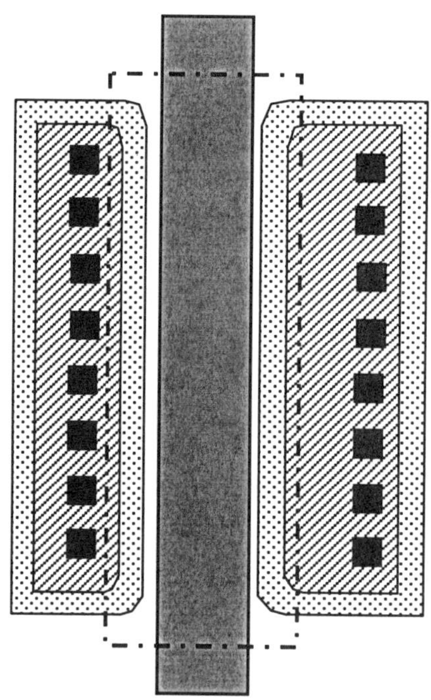

Figure 7.11 A layout for a thick-gate NMOS ESD protection structure features rounded the source and drain diffusions for uniform current and heat distribution.

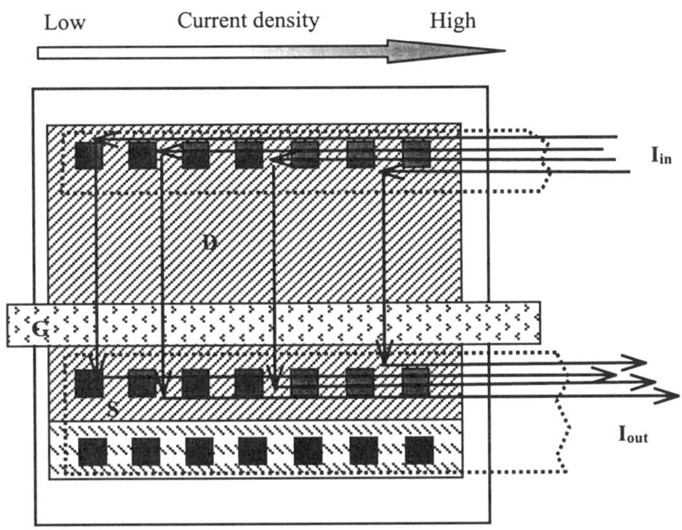

(a)

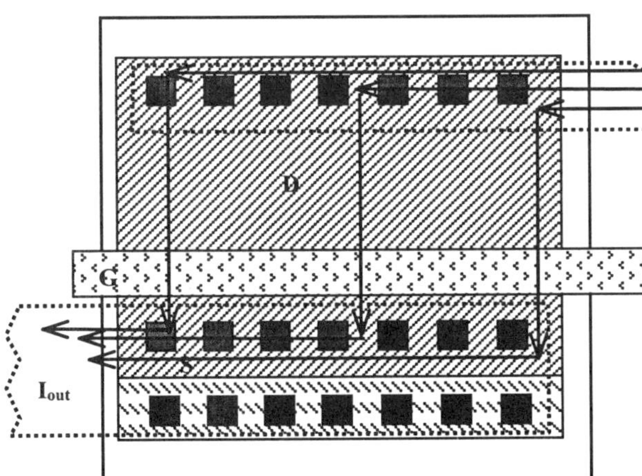

(b)

Figure 7.12 An anti-parallel metal routing (a) may cause current crowding at right, while a parallel metal layout (b) ensures an even current distribution.

patterns, such as, B-DGSGD-B-DGSGD-B, BSGDGSGDGSB and BSGDGSBSGDGSB [10]. Figure 7.10 illustrates such an optimised layout pattern for a multiple-finger NMOS ESD protection structure. For the same uniform current distribution consideration, source and drain diffusion fingers are often smoothed to avoid current crowding and local overheating as shown in figure 7.11 [11].

Contact placement and metal interconnect routing play vital roles in achieving uniform current and heat distribution. Figure 7.12 (a) illustrates a *bad* anti-parallel metal line layout, where current crowding is expected at the right, resulting in thermal damage there. A *good* parallel metal line routing scheme is shown in Figure 7.12 (b) that leads to a much even current distribution. A much better metal routing scheme is given in Figure 7.13, where metal lines follow the current flow direction. Of course, it may not always be possible to routing metal this way. Nevertheless, a careless contact

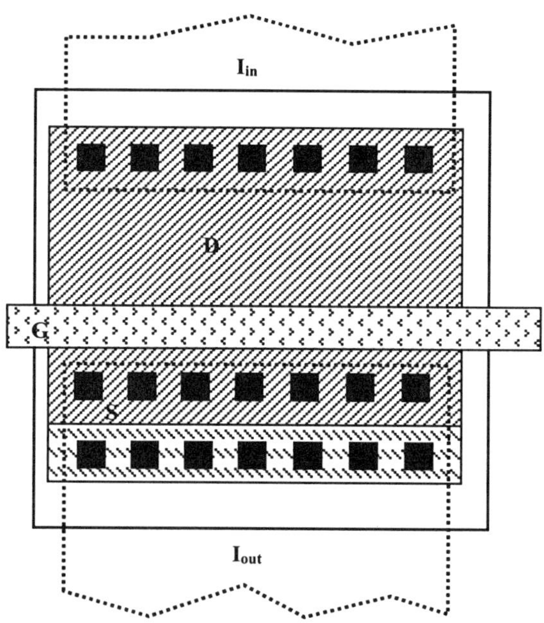

Figure 7.13 Routing metal lines parallel with the ESD current flow direction further improves ESD protection level.

planning and metal routing practice as shown in Figure 7.14 must not be used at all. Because an un-even contact placement combined with an anti-parallel metal routing will certainly guarantee a double chance of overheating at the right. When multiple metal layers are used for ESD protection structure metal interconnects, careful planning for vias are as important as for the contacts in order to ensure uniform pick-up of ESD currents from lower metal to higher metal. As shown in Figure 7.15, the top portion demonstrates an even contact-to-via placement that gives a uniform current pick-up. However, the lower portion shows a very bad contact-to-via planning, which will lead to current crowding and thermal failure in vias.

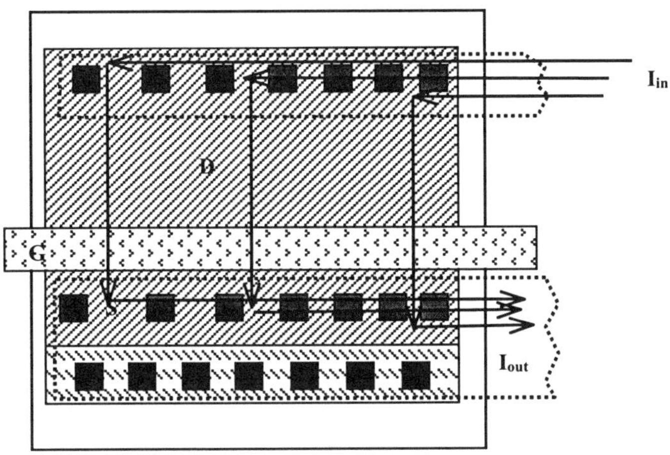

Figure 7.14 A poor layout scheme with un-even contact placement and anti-parallel metal routing may double the overheating effect.

A lot more good layout tricks can be used to achieve better ESD protection performance for the same ESD protection structures if cares are exercised. The benefits of a well-thought layout include more uniform heat distribution, maximized ESDV level and reduced protection device size. For example, for a complementary MOSFET-triggered SCR protection scheme designed for output buffers as shown in Figure 7.16, a ggNMOS-triggered SCR and a ggPMOS-triggered SCR, shown in Figure 7.17, are used to lower the trigger voltage. If the MOS-SCR units are created separately in layout, a large size is expected due to individual guard-rings and spacing requirement. A good layout practice is to place the MOS-SCR ESD protection structures and the output buffer MOSFET transistors being protected inside the same guard-ring region as illustrated in Figure 7.18 and Figure 7.19 [12]. This layout scheme not only saves silicon area, but also ensures better MOSFET triggering result. Obviously, the channel length of the MOSFET transistors, L_2, should be designed somewhat narrower than that of the output buffer MOSFET transistors being protected, L_1, to eliminate possible competing turn-on of the output buffer MOSFET transistor during an ESD event. All the above suggest that a little mind work may lead to substantial improvement in ESD protection circuit design.

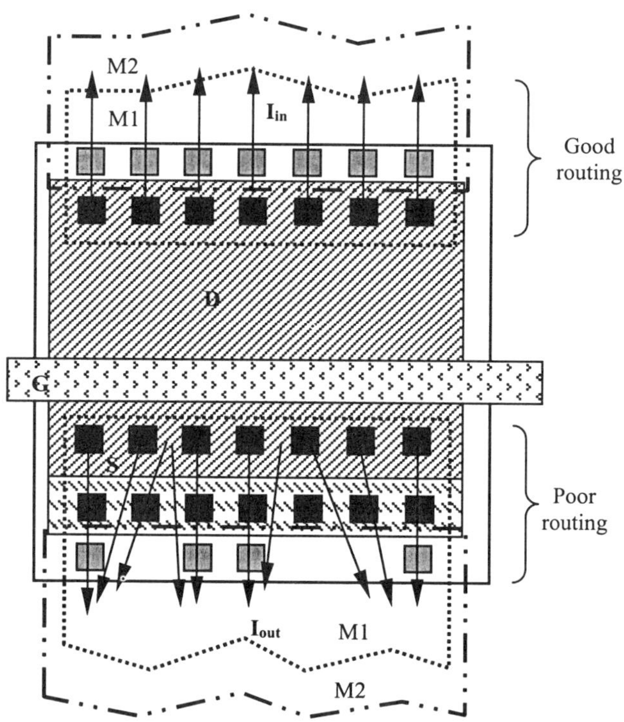

Figure 7.15 a good contact-via placement (upper portion) ensures uniform current pick-up from contacts to vias; while a bad planning (lower portion) leads to localized overheating.

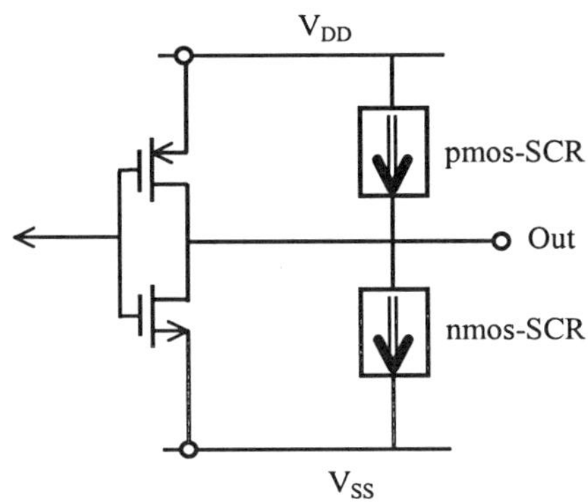

Figure 7.16 A complementary MOS-triggered SCR output ESD protection scheme [12].

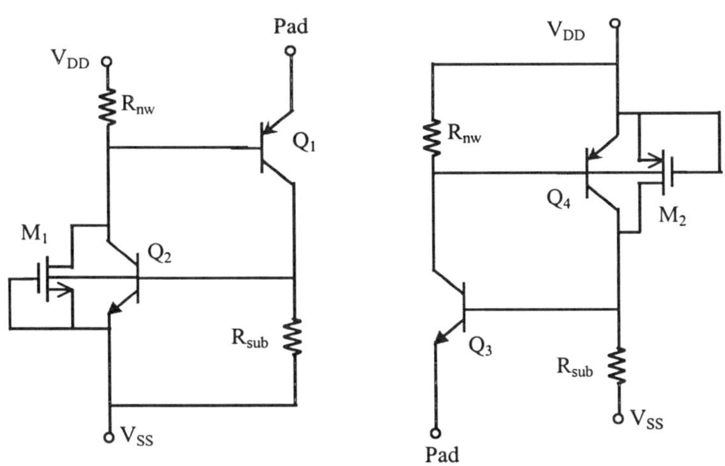

(a) (b)

Figure 7.17 A NMOS-triggered SCR (a) and a PMOS-triggered SCR (b) ESD protection structures [12] (*Reproduced here by kind permission of IEEE and authors*).

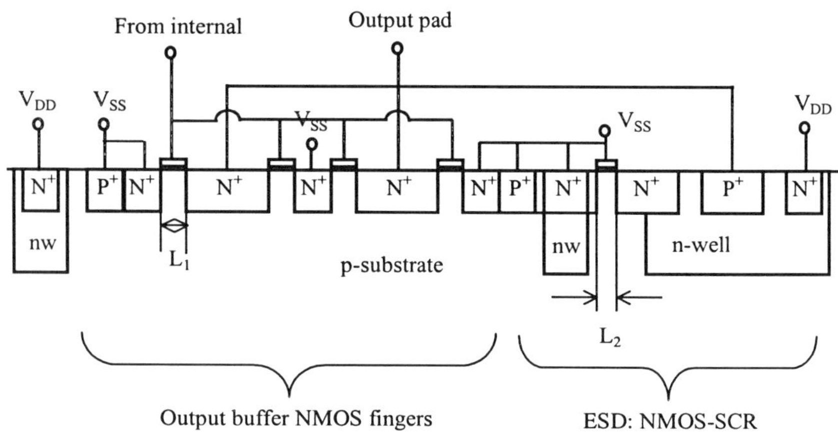

Figure 7.18 A combined output buffer NMOS ladder and NMOS-SCR ESD protection structure [12] (*Reproduced here by kind permission of IEEE and authors*).

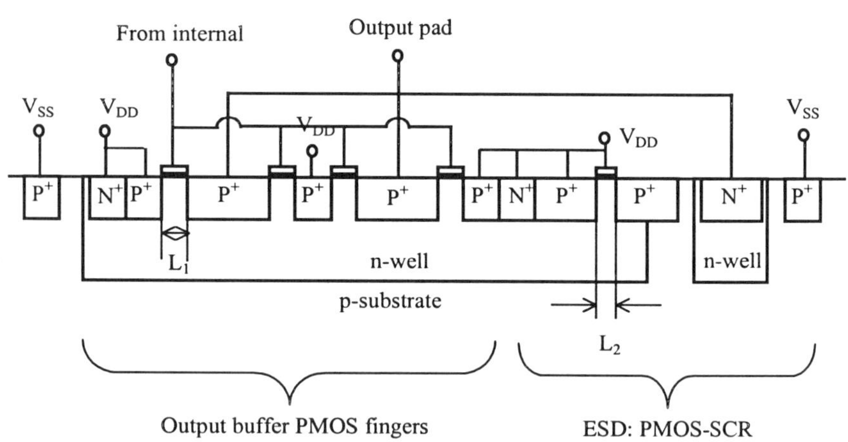

Figure 7.19 A merged layout for an output PMOS buffer transistor and a PMOS-SCR ESD protection structure [12] (*Reproduced here by kind permission of IEEE and authors*).

7.4. ADVANCED LAYOUT DESIGN CONCEPTS

Although bells and whistles in layout design benefit ESD protection performance noticeably, these tricks are just the final touches in a master artwork. As IC technologies continuously advance and IC chips become more sophisticated, novel ESD protection design featuring high current handling capacity, small size and low parasitic effect is a highly desirable and challenging design task. Critical concerns in this regard include to achieve adequate ESD protection level with less silicon area and lower ESD-induced parasitic effect; to realize super high ESD robustness with reasonable size; and to accommodate the protection needs for ultra fast ESD transients, e.g., CDM and IEC mode ESD pulses. This is where a circuit designer can demonstrate his/her talents. Other than to develop novel structures, innovative layout design plays a big role in developing novel ESD protection solutions. A few design examples are discussed in this section.

As an alternative to the commonly used ladder NMOS ESD protection structure in CMOS technologies that becomes less efficient as IC feature size continuously shrinks, Figure 7.20 shows an NMOS ESD protection structure with a waffle layout pattern. This cell-based structure has an advantage of uniform current distribution, which leads to better ESD protection. This waffle structure demonstrated better ESD protection performance in CMOS technology with LDD and silicidation features, where the sharp cell corners were smoothed to suppress localized electrical field overstress [13].

In another design using complementary ggNMOS/ggPMOS ESD protection scheme, octagonal cells are proposed with the same goal of achieving uniform current distribution, hence, higher ESDV result. In addition, this octagonal structure is much more friendlier to pad layout planning as opposed to a bulky ladder structure. Figure 7.21 illustrates such a cell layout design, where P^+ body pick-ups are evenly distributed across the ESD protection area to improve current uniformity. Figure 7.22 shows a micro-photo for such an design implemented in a 0.6μm CMOS technology, which demonstrated an improvement of ~40% in unit ESDV result for HBM ESD zapping over a traditional multiple-finger MOSFET ESD protection design as per the report [14].

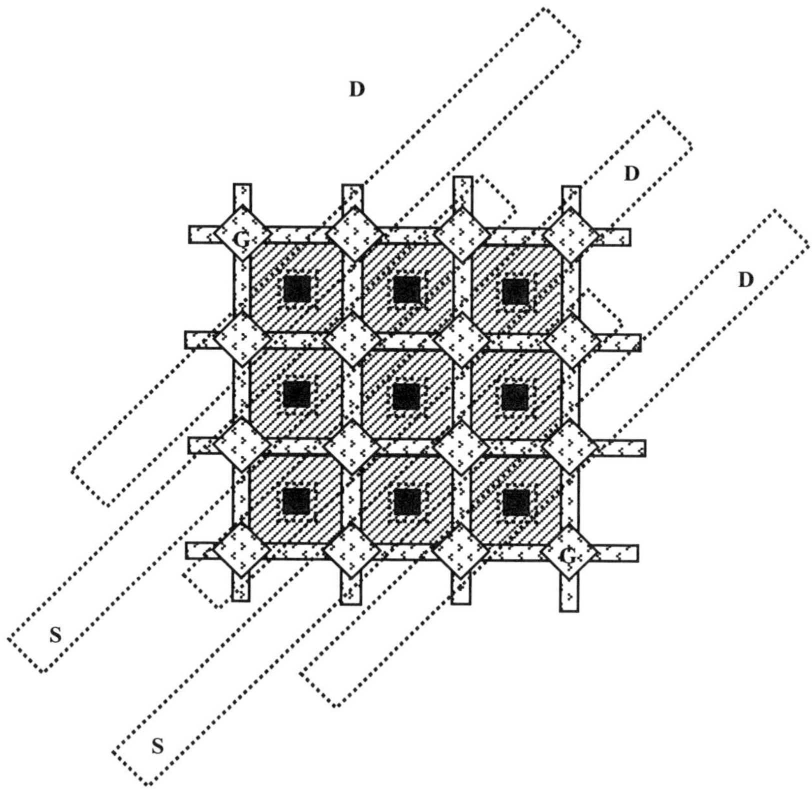

Figure 7.20 A waffle structure of NMOS ESD protection design features uniform current distribution and delivers good ESD protection result in an LDD and silicided CMOS technology [13] (*Reproduced here by kind permission of the ESD association and authors*).

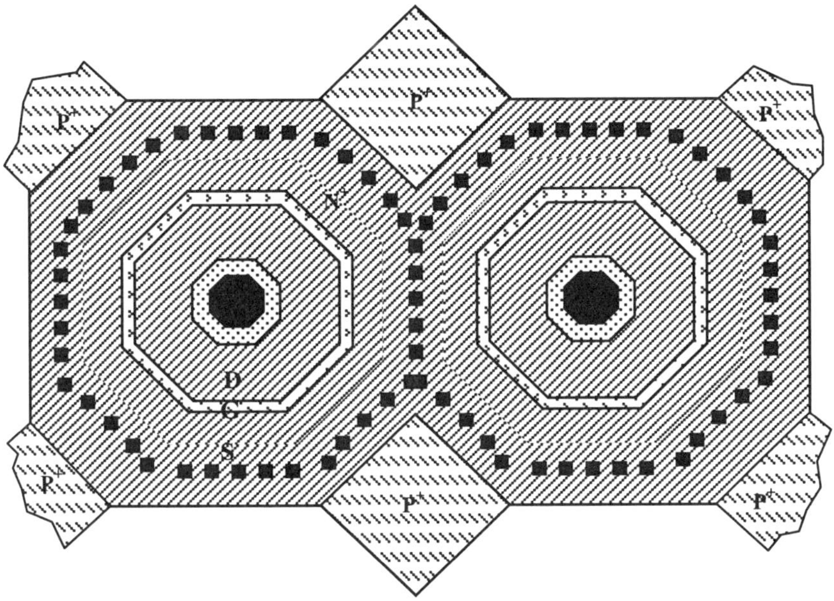

Figure 7.21 A octagonal cell layout design for MOSFET ESD protection structure [14] (*Reproduced here by kind permission of IEEE and authors*).

Figure 7.22 Amicro-photo for a complementary NMOS/PMOS ESD protection using the octagonal cells depicted in Figure 7.21 [14] (*Reproduced here by kind permission of IEEE and authors*).

Following the similar concept, a cell-based design for the dual-direction SCR type ESD protection structure, as depicted in Figure 5.8, is devised to ensure uniform current distribution and hence, better unit ESDV result. Figure 7.23 shows the layout of a single cell consisting of a centre P^+ diffusion as one terminal, an N^+ diffusion ring as another terminal, and an outer n-well guard-ring as the boundary between cells. The dual-direction SCR ESD structure is depicted clearly in Figure 7.24, which is a cross-section view along the AA' cut-line for two adjacent cells. It is readily to understand that each side of the cell contributes to ESD current discharging, therefore, leads to higher unit ESDV result. Figure 7.25 is a micro-photo of such an ESD protection structure implemented in a 0.6μm BiCMOS technology that delivers a high unit ESDV value of ~ 1.66V/μm² [15]. It is worth to note that one ought to take care of the possible accidental turn-on problem in using SCR ESD protection structures, as well as other ESD protection structures, such as NMOS. Considering the fact that most of the ESD protection structures rely on turn-on of the parasitic BJT transistors for

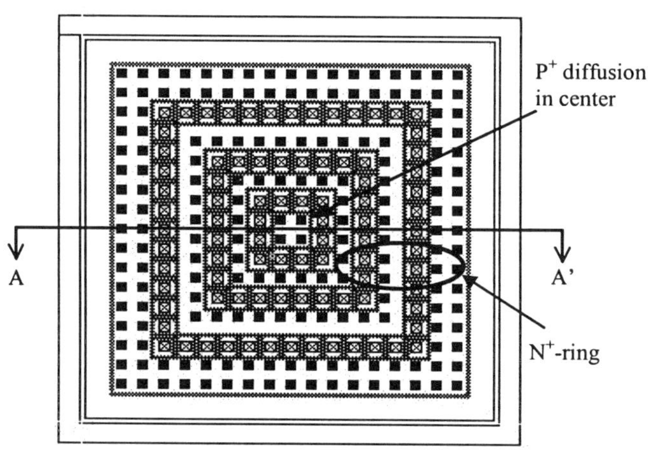

Figure 7.23 One cell for a cell-based dual-direction SCR type ESD protection structure has a centre P^+ diffusion as one terminal, an N^+ diffusion ring as another terminal, and a n-well guard-ring [15] (*Reproduced here by kind permission of IEEE and authors*).

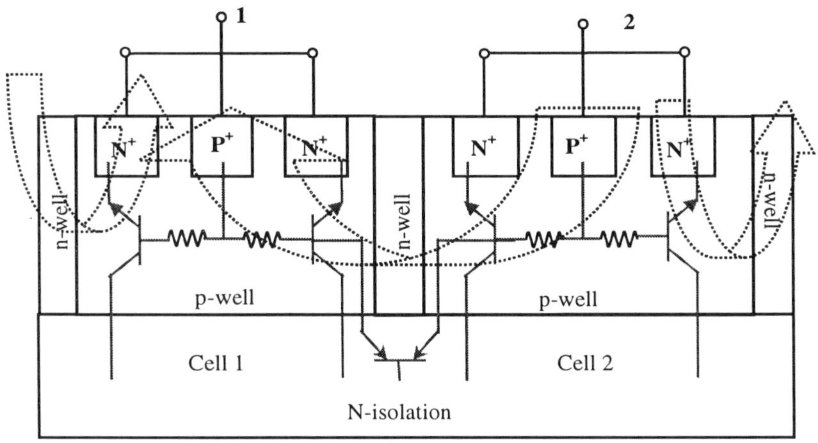

Figure 7.24 A cross-section of a dual-direction SCR ESD protection structure formed between two adjacent cells along the AA' cut-line shown in Figure 7.23 [15] (*Reproduced here by kind permission of IEEE and authors*).

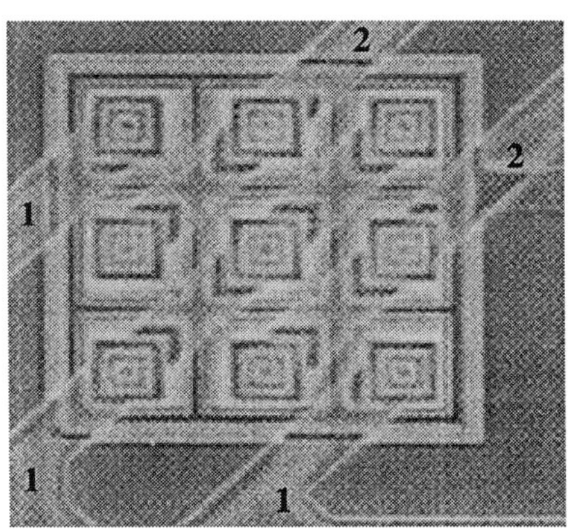

Figure 7.25, A micro-photo of the cell-based dual-direction SCR-type ESD protection structure in BiCMOS [15] (*Reproduced here by kind permission of IEEE and authors*).

triggering that is fired by substrate currents, an ESD protection structure should be properly isolated by guard-rings and be placed away from any large current generating device on a chip. It is reported that substantial substrate current coupling from neighbouring sub-circuits, such as a charge pump unit, can cause unwanted triggering of the ESD protection structure under normal circuit operation [16].

Provided a working ESD protection structure is available, one main issue left for a circuit designer is where to place it on a chip. Often, an ESD protection structure is too large to make a good pad plan, particularly in high-pin-count design for VDSM applications, where silicon consumption by ESD protection units becomes a real cost factor. To address this concern, it would be nice to have a bounding pad oriented ESD protection design that can be placed underneath or surrounding a pad. Such pad-oriented ESD protection structure requires very careful layout design and, sometimes, violation of existing design rules. Nevertheless, the benefits gained could be significant. For example, an under-pad design of a PNP diode ESD protection device, as shown in Figure 7.26, is reported for an 8000-pad chip in a regular CMOS technology [17]. The P^+/n-well diode is in a PNP format as shown in Figure 7.27 for a three-metal CMOS process. Careful contact-via planning and metal routing is essential to ensure correct operation of the ESD protection structure as illustrated in Figure 7.27 and Figure 7.28. One can easily image the kind of saving of silicon asset resulted from putting the ESD protection units underneath bounding pads given the large number of pads of the chip. In another example, a pad-oriented ESD protection design

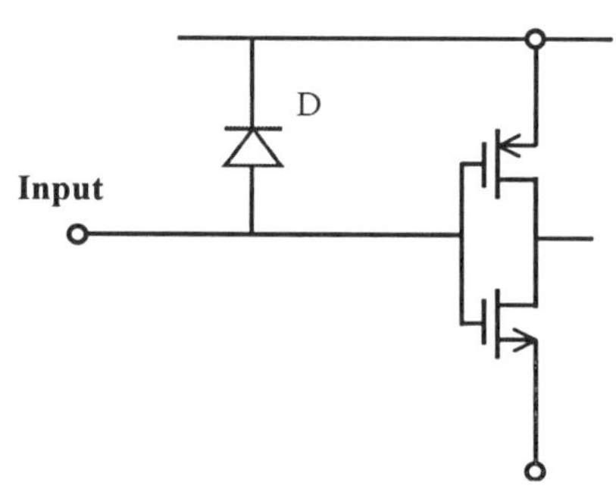

Figure 7.26 A PNP diode ESD protection scheme is used for a CMOS technology.

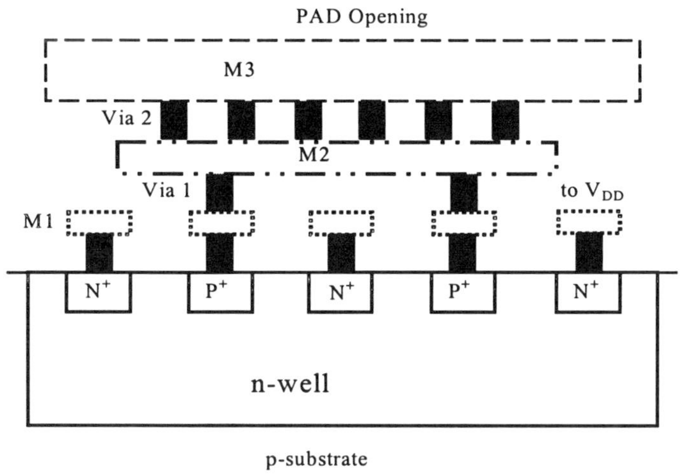

Figure 7.27 A cross-section, along A-A' cut-line in Figure 7.28, for the under-pad diode ESD protection [17] (*Reproduced here by kind permission of the ESD Association and authors*).

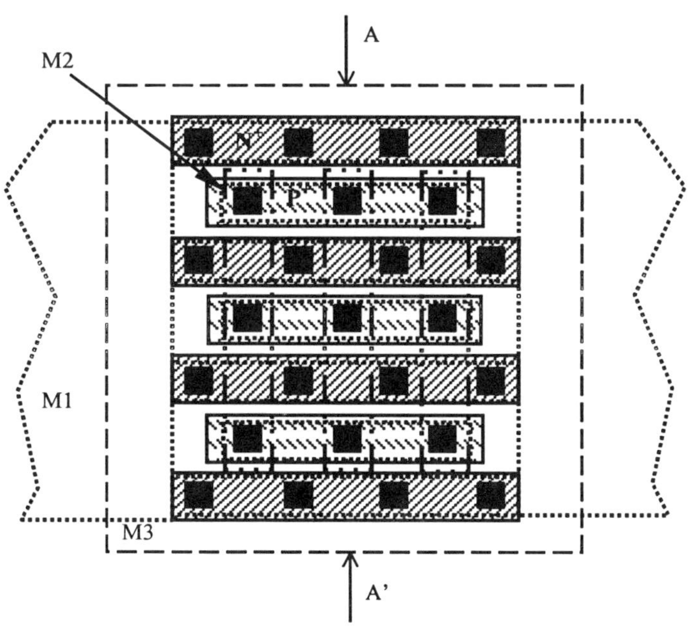

Figure 7.28 A top view of the under-pad diode ESD protection unit shown in Figure 7.27 [17] (*Reproduced here by kind permission of the ESD Association and authors*).

corresponding to the all-in-one ESD protection structure depicted in Figure 5.11 is shown in Figure 7.29, where special layout divides the whole structure into four units to discharge ESD transients of all modes, i.e., PD, ND, PS, NS and DS. Figure 7.30 shows a micro-photo for the design implemented in a BiCMOS technology [18]. To further improve the current

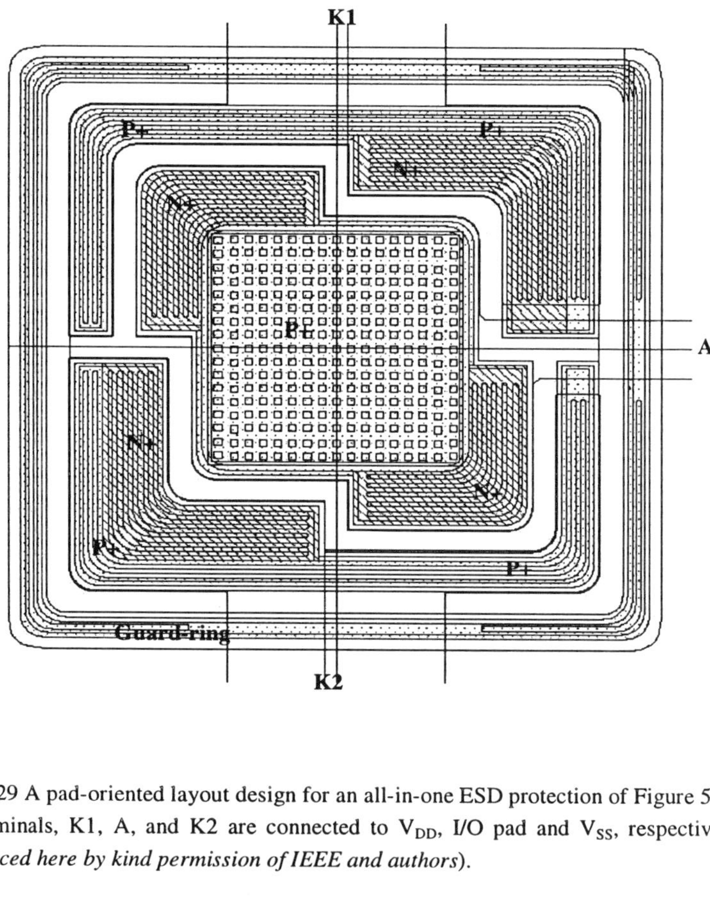

Figure 7.29 A pad-oriented layout design for an all-in-one ESD protection of Figure 5.11. The three terminals, K1, A, and K2 are connected to V_{DD}, I/O pad and V_{SS}, respectively [18] (*Reproduced here by kind permission of IEEE and authors*).

distribution, a circular design is shown in Figure 7.31, where the multiple-mode ESD protection operation can be readily understood from the cross-sections along the diagonal cut-lines, 0-1, 0-2, 0-3, and 0-4 for the SCR devices operating under PS, NS, PD and ND ESD transient modes, respectively [19].

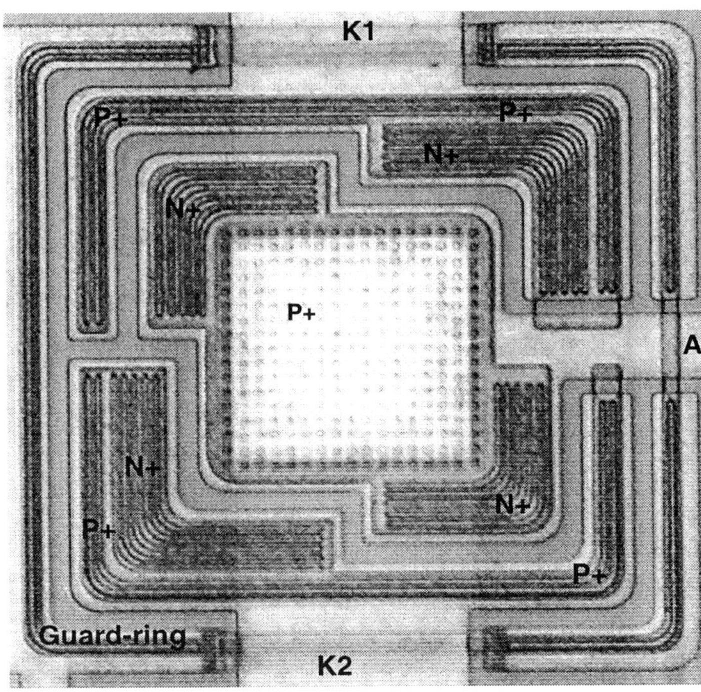

Figure 7.30 A micro-photo for the pad-oriented all-in-one ESD protection structure of Figure 7.29 in BiCMOS [18] (*Reproduced here by kind permission of IEEE and authors*).

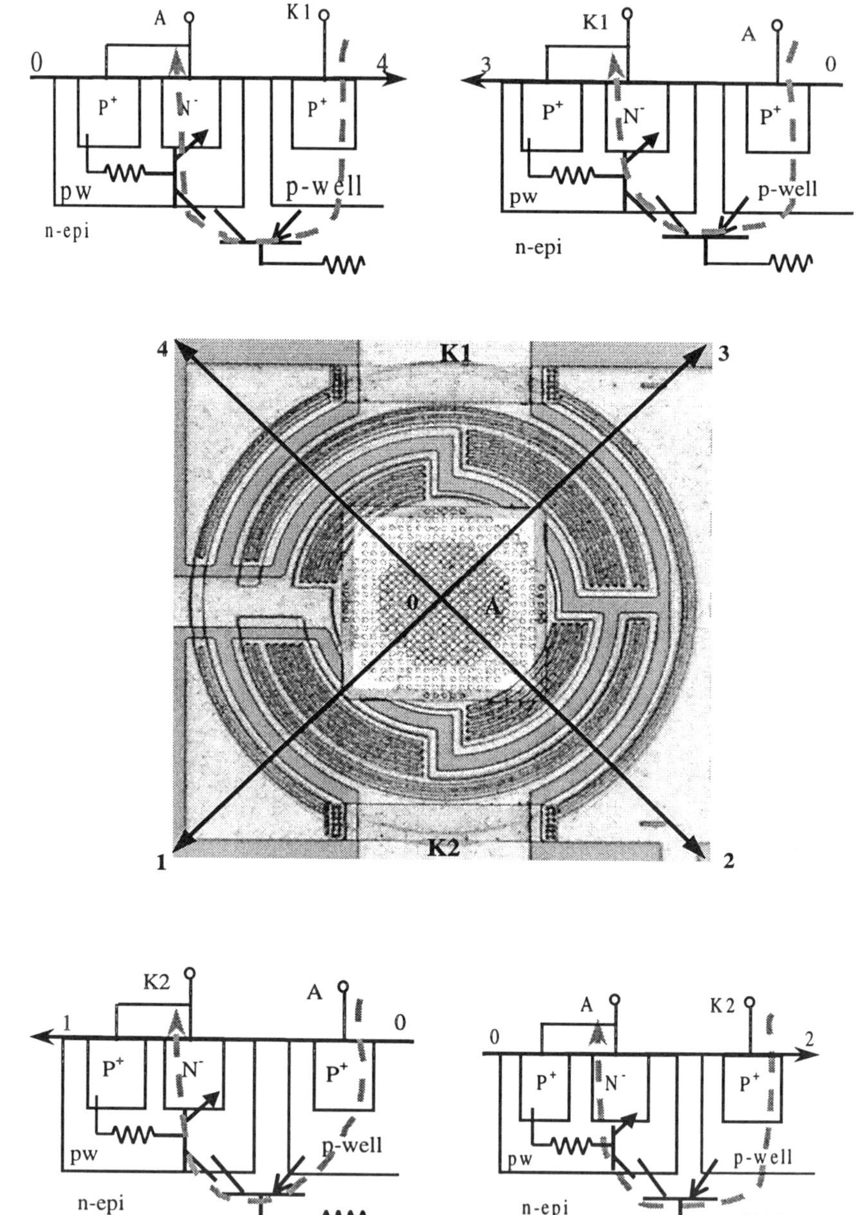

Figure 7.31 A pad-oriented circular design for an all-in-one ESD protection structure in BiCMOS [19].

A pad-oriented ground-gate MOSFET ESD protection design is illustrated in Figure 7.32 [4]. MOSFET ESD protection structure is widely used in CMOS technologies because it is one of the most understood and stable ESD protection elements, for which many device models are available for ESD circuit simulation. The complementary MOSFET ESD protection scheme shown in Figure 7.32 consists of one ggPMOS device connected from the I/O pad to V_{DD} and a ggNMOS ESD protection unit connected from the pad to V_{SS}, hence, provides complete ESD protection for the I/O pad. The MOSFET ESD protection devices can be either placed underneath or surrounding the pad being protected. The insets show an enlarged view of the DGSB layout pattern and its cross-section. The design features optimal DCGS spacing, minimum SCGS spacing and guard-rings. Since the layout follows the peripheral of the pad, it provides a large effective ESD discharging channel that makes high ESDV result possible.

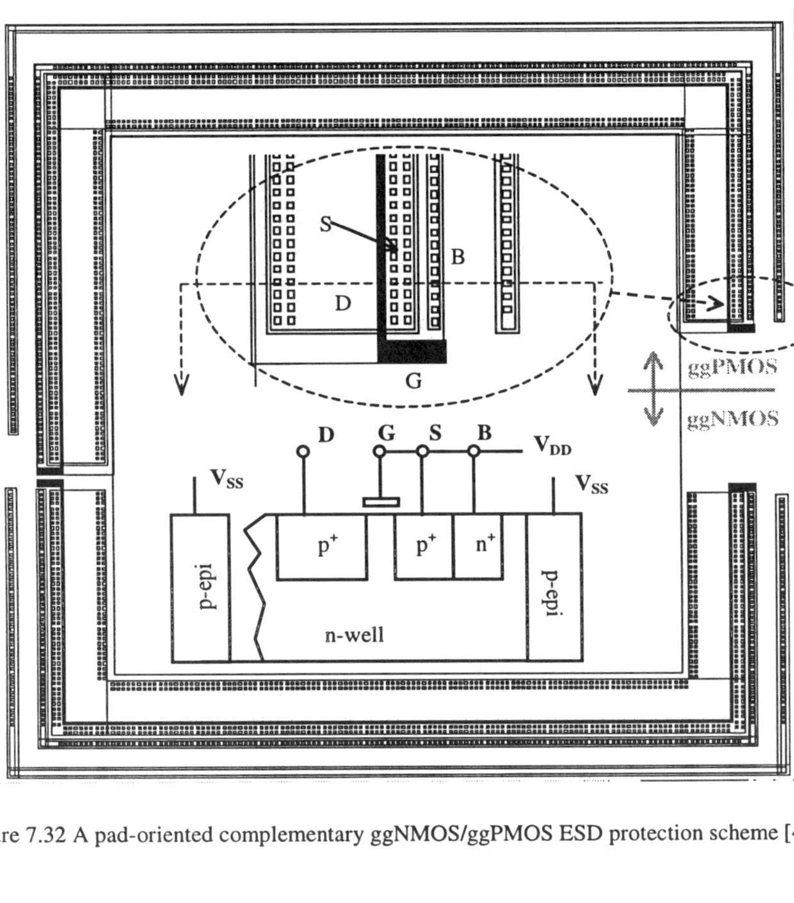

Figure 7.32 A pad-oriented complementary ggNMOS/ggPMOS ESD protection scheme [4].

Figure 7.33 shows a similar pad-oriented ggNMOS-triggered SCR ESD protection structure [4] as depicted by Figure 4.11. The insets illustrate an enlarged layout view and its cross-section. Critical spacing such as the ggNMOS channel length, L_1, and N^+ to n-well spacing, L_2, need to be optimised for better ESD protection operation.

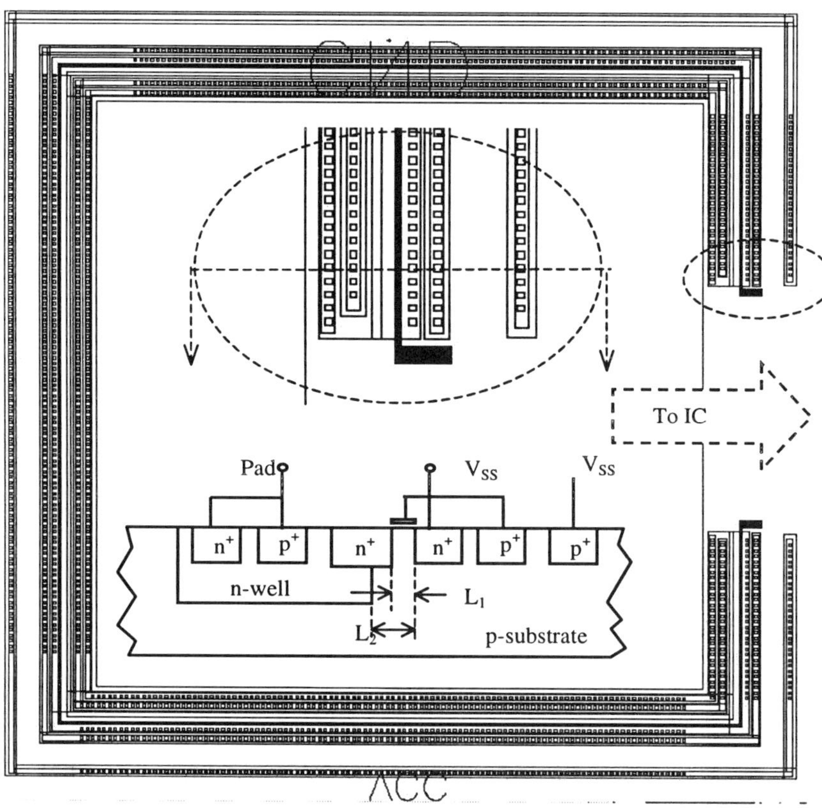

Figure 7.33 A pad-oriented ggNMOS-triggered SCR ESD protection structure [4].

7.5. TECHNOLOGY SCALING *vs.* ESD PROTECTION

IC process technologies play big roles in ESD protection design and performance. Particularly, the CMOS technology scaling, the main driving force of improving transistor performance, has tremendous impacts on ESD protection design [20, 21]. Typically, following CMOS scaling theory, n-well and epitaxial layer scaling lead to degradation of ESD protection. A study using P^+/n-well diode ESD protection device as the benchmark for technologies scaling from 1.2μm to 0.25μm with critical process parameters listed in Table 7.1 indicates that, using similar scaling factors for diode ESD protection devices except for maintaining the same diode width, unit ESD protection level per device width decrease monotonically [20]. However,

Table 7.1 A summary for a technology scaling ~ ESD protection study using a diode ESD protection device as the benchmark [20] (*Reproduced here by kind permission of the ESD Association and authors*).

Feature size (μm)	1.2	0.7	0.5	0.25
t_{ox} (Å)	235	150	135	70
epi (μm)	12	2.5	2.0	2.0
Well depth (μm)	4.0	1.4	1.2	0.9
L_{eff} (μm)	1.0	0.7	0.5	0.25
P^+ diode (μm)	4.0	2.0	2.0	1.5
Ti:Si	No	Yes	Yes	Yes
STI	No	No	Yes	Yes
ESD diode length (μm)	74	60	43	30
ESD diode width (μm)	**120**	**120**	**120**	**120**
ESDV (kV)	**9.0**	**6.5**	**4.3**	**3.0**

things are not that hopeless as indicated by the ESDV values given in Table 7.1. Another study for NMOS ESD protection structures under technology scaling, as summarized in Table 7.2, demonstrates that positive factors like reduced NMOS channel length could compensate the negative effects such as shallower junction depth. Consequently, overall unit ESD protection performance may be maintained at a similar level of ~ 8V/μm to the 0.25μm node for LDD silicided CMOS technologies [21]. Indeed, with careful design and innovative structures, ESD protection has not yet been in the downhill ride, at least down to the 0.25μm node.

Table 7.2 A summary of technology scaling parameters over which NMOS ESD protection performance was studied [21] (*Reproduced here by kind permission of IEEE and authors*).

Feature size (μm)	3.0	2.0	1.5	1.0	0.8	0.5	0.35	0.25
t_{ox} (Å)	500	400	300	200	150	100	70	50
Junction depth (μm)	0.8	0.5	0.4	0.35	0.3	0.25	0.2	0.15
LDD	No	No	Y	Y	Y	Y	Y	Y
Salicide	No	No	No	Y	Y	Y	Y	Y

7.6. NEW TECHNOLOGY *vs.* ESD PROTECTION

New process technology features are constantly being introduced into mainstream CMOS IC technologies to boost transistor performance, such as, speed, current driving capacity, etc. These new process features may influence ESD protection performance either negatively or positively. One of the most widely used process features in CMOS technologies is LDD implantation, which aims to suppress the troublesome hot-carrier effect and boost device lifetime. Unfortunately, this nice LDD feature degrades ESD protection dramatically [22]. One of the main issues of the LDD in ESD operation is associated the strong electrical field at the shallow LDD junction. Several fixing measures are used to resolve the LDD problem. First, one can use a deeper double-doped-drain (DDD) to replace the LDD. The disadvantage with DDD is that it alters the baseline CMOS transistor doping profiles negatively. Second, an extra LDD-blocking mask is widely used to boost ESD protection in an LDD CMOS process, where the ESD protection structure regions are blocked from the LDD implantation, as illustrated in Figure 7.34. Of course, the extra mask and associated process steps increase the costs. Third, an extra implantation step, called ESD implant, may be used in the ESD protection structure region only to create a deeper, DDD-like junctions in MOSFET ESD protection devices, while maintaining the desired LDD feature in all the regular MOSFET transistors [23].

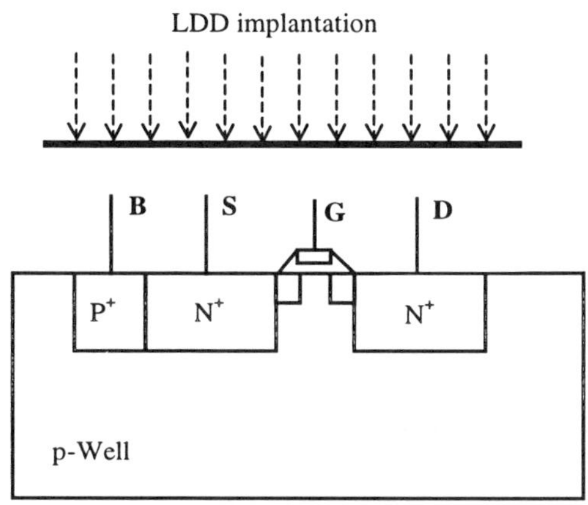

Figure 7.34 An LDD-block mask can be used to prevent LDD implantation from being injected into ESD protection structures.

Another commonly used new process feature in CMOS technologies is silicidation of silicon source/drain and poly-silicon gate in order to reduce the sheet resistance in source, drain and gate, as illustrated in Figure 7.35. This silicide, or salicide (self-aligned silicide), layer will certainly cause ESD protection problem. Recall that the main reason for using a sufficient DCGS spacing is to introduce a drain extension resistance, i.e., ballasting resistance, to realize uniform turn-on of multiple-finger MOSFET ESD protection transistors. With the introduction of silicide layer, the valuable drain ballasting resistance decreases to a negligible value. Reducing silicide layer thickness, *d*, will bring back the ballasting resistance to some extent, hence benefits ESD protection performance [24], which may not be preferred for regular transistor operation. One fixing measure to this problem is to introduce one extra silicide-blocking mask to prevent any silicidation

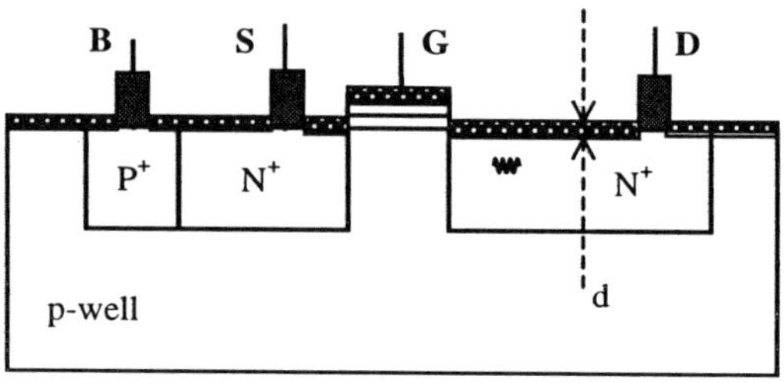

Figure 7.35 Silicidation causes diminishing of the desired drain extension resistance, resulting in degradation in ESD protection performance.

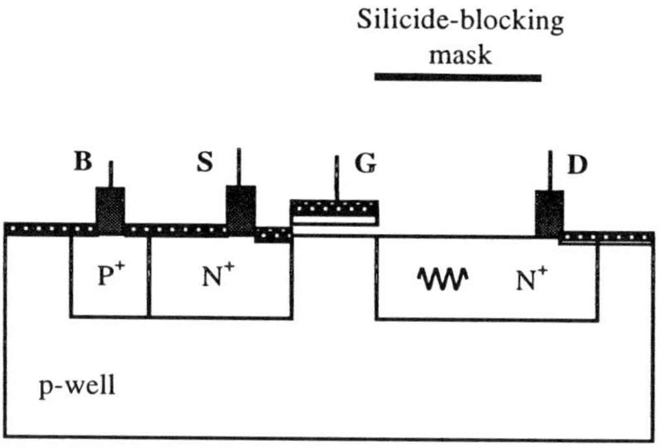

Figure 7.36 A silicide-blocking mask can be used to prevent silicidation in the drain region to prevent degradation in ESD protection performance.

within the ESD protection structure regions, as shown in Figure 7.36. The disadvantage is, again, the extra costs associated with the extra process steps. One fairly smart measure to address the silicidation problem is to use a dummy, floating gate as the silicide-blocking mask, therefore, to eliminate the added costs of using an extra silicide-blocking mask. Well, if one is willing to pay for this extra cost in exchange of substantial improvement in ESD protection performance, an extra p-base diffusion implantation may be added to CMOS process to realize a robust vertical NPN bipolar protection transistor. One such design using NPN diode ESD protection is shown in Figure 7.37, where a floating gate serves as a silicide-blocking mask and a p-base diffusion is used to form a superb vertical NPN transistor in the CMOS process for better ESD protection [25]. Similarly, more after-qualification fixing measures can be used to boost ESD protection performance, or, simply to bring back to the original ESD protection level degraded by some nice process features. One major disadvantage in relying on such fixing measures is that added costs occur. A good approach shall be to include the ESD protection design into the initial process development phases. Very often, trade-offs can be made to realize super ESD protection values without

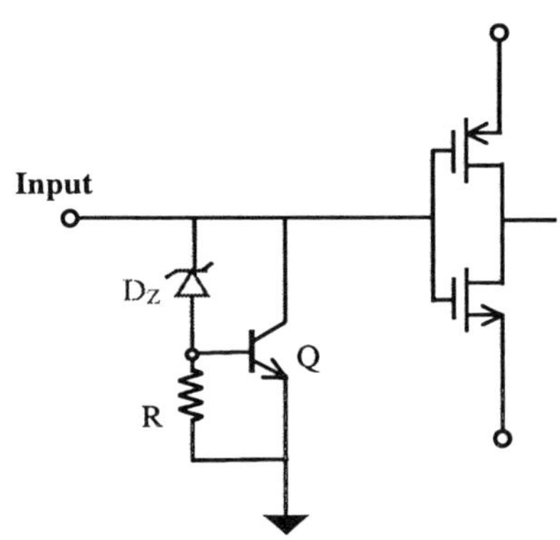

(a)

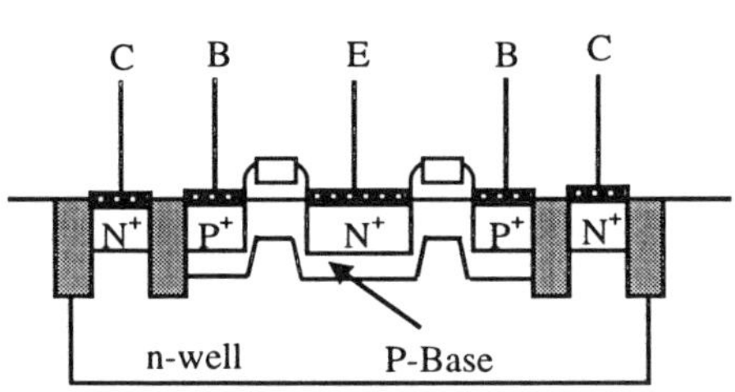

(b)

Figure 7.37 A dummy gate is used as the silicide-blocking mask and a p-base diffusion is introduced to make a vertical NPN device in a CMOS technology. A diode ESD protection structure is developed in this technology [25] (*Reproduced here by kind permission of IEEE and authors*).

suffering much in regular transistor operation. For example, a relative thinner silicide layer may be accepted for transistor operation. LDD dopant and dose may be selected to meet the needs for both transistors and ESD protection structures. One can also optimise the doping profiles and thickness of well and epitaxial layers. After all, one may surprisingly find out that there are not always quantitative reasons for most specifications of process receipts, such as, an LDD implant dose of $5x10^{15}$ /cm^{-2} or $5.5x10^{15}$ cm^{-2}. However, that minor variation in LDD doping may make noticeable difference in ESD protection performance at no extra cost at all.

Process technology associated with contact, via and metal interconnect is yet another big issue in ESD protection circuit design. Inappropriate layout may lead to early ESD failure. One has to make sure the number of contact and vias as well as their placement can survive the targeted ESD transient level. In terms of metal interconnect, the routing should be optimised and metal width should be sufficient. Ideally, the total numbers of contacts and vias, and the width of metal lines should be determined based upon their maximum current handling capacities. However, existing process design rules normally do not provide such reliability data for ESD stresses. The normal DC and AC electrical stress data do not apply to ESD pulsing situation. For this reason, common practice in ESD metal design is experience-based, or, is a kind of blind design. One widely used design rule is to use 20μm wide metal lines or as wide as possible a metal line for the ESD protection structure. This is certainly not reasonable considering technology scaling. While the conservative metal design rule may eliminate metal blowout under ESD pulses, excessive metal coverage used in ESD protection structures will produce extra metal-induced parasitic capacitance, which may affect high-speed circuit operation noticeably [26]. A rational metal design for ESD protection structure using simulation is essential to full-chip ESD protection circuit design. Although special CAD tool is not yet available for such simulation, research shows using existing TCAD tool to include metal interconnect into to ESD simulation can be beneficial, which will be discussed in Chapter 8.

7.7. ESD PROTECTION FOR SOI AND SiGe

One of the very promising technologies for future high-performance IC chips is SOI technology. SOI is gradually moving into the mainstream IC technology club. Particularly, SOI CMOS technology allows almost direct mapping of regular CMOS technologies into the SOI process flows. While SOI offers significant advantages over bulk silicon technologies in transistor

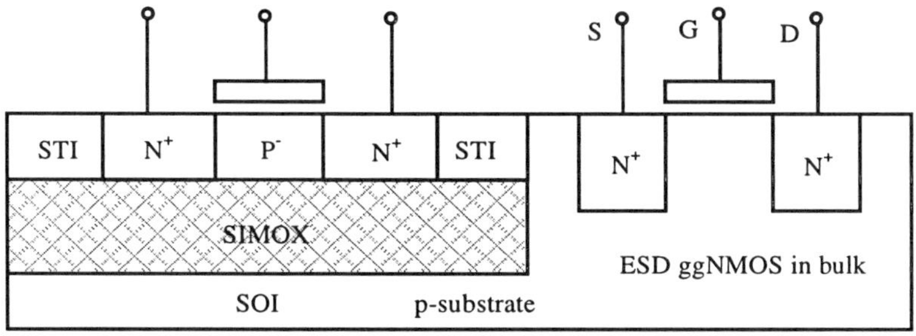

Figure 7.38 A combined SOI-bulk ESD protection scheme for SOI ICs places the MOSFET ESD protection device inside bulk silicon for better ESD robustness [27] (*Reproduced here by kind permission of IEEE and authors*).

and circuit operation, it poses extra challenges to ESD protection circuit design for SOI ICs. One major problem is associated with its poor thermal dissipation property in SOI structure, because the very thin SOI Si film is covered by poor thermal conducting materials, i.e., oxide layer on the top and insulating oxide layer at the bottom. It is reported that ESD protection performance for SOI is only ~55% of that achieved in bulk silicon for MOSFET ESD protection structures [27]. For this reason, One ESD protection solution proposed for SOI applications is to build ESD protection devices in the bulk Si region on the same chip, as shown in Figure 7.38 [27]. Of course, this SOI/bulk combination is only possible if the SOI technology uses SIMOX (separation by implanted oxygen) or similar techniques. Many other ESD protection structures are also available for SOI technologies. Figure 7.39 shows a SOI version of ggNMOS ESD protection structure, where a lateral NPN operates under the positive ESD pulse while a body/drain diode provides protection against the negative ESD transient, provided that a body connection is available [28, 29]. Its I-V operation is similar to a ggNMOS device in bulk silicon except that two snapback points are observed as illustrated in Figure 7.40 [29]. The first snapback action still corresponds to the triggering of the lateral NPN transistor as in the bulk NMOSFET case. It is suggested that, after the first snapback, electron current flows in sub-surface region and forms a floating neutral zone in the depleted region. An increase in bipolar current gain occurs as the potential of the neutral zone builds up to sufficiently higher than that of the source. The

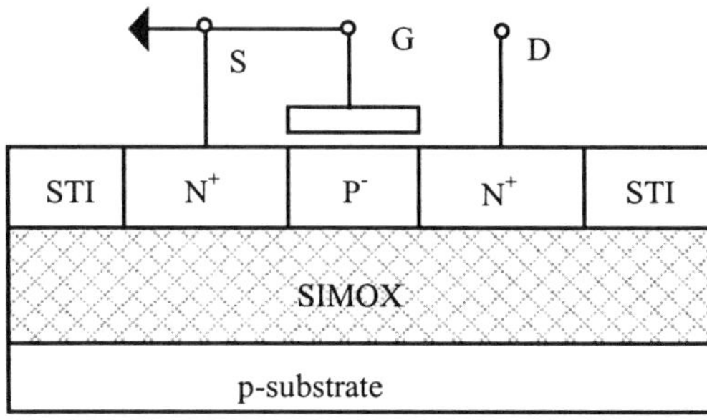

(a)

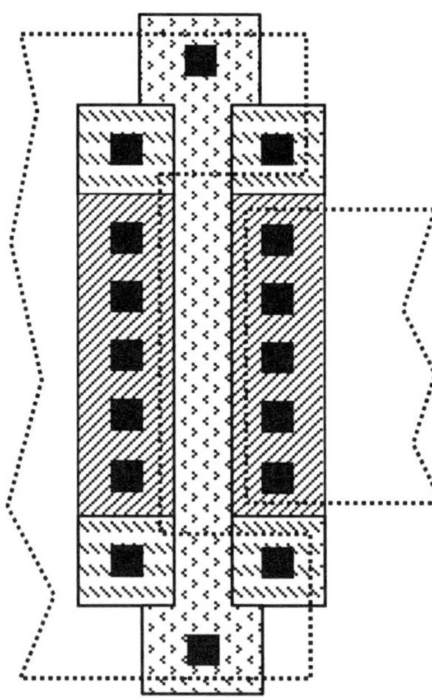

(b)

Figure 7.39 A ggNMOS ESD protection structure for SOI, where body connection is available.

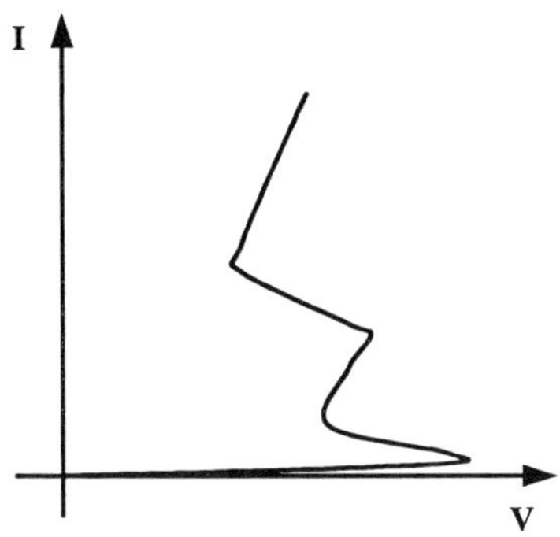

Figure 7.40 A double-snapback I-V characteristic is observed for a ggNMOS ESD protection structure with a body connection for SOI.

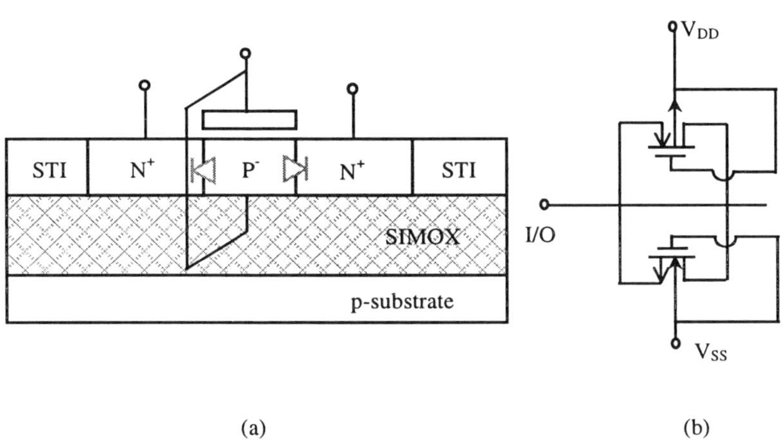

Figure 7.41 A double-diode ESD protection scheme for SOI uses two diodes, i.e., the source and drain diodes, to discharge ESD transients, where body and gate are connected together [28] (*Reproduced here by kind permission of Elsevier Science and authors*).

second, lower snapback action then takes place in this high-gain bipolar device [29]. Although further study is needed to thoroughly understand this double-snapback phenomenon, benefit can be expected out of the deeper second holding voltage for ESD protection. Various diode ESD protection structures are also proposed for SOI applications with some examples depicted in Figure 7.41, Figure 7.42 and Figure 7.43 [28].

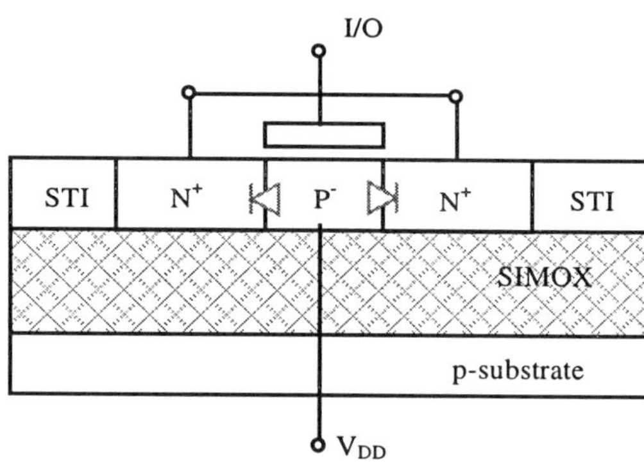

Figure 7.42 A grounded-gate diode ESD protection structure for SOI [28] (*Reproduced here by kind permission of Elsevier Science and authors*).

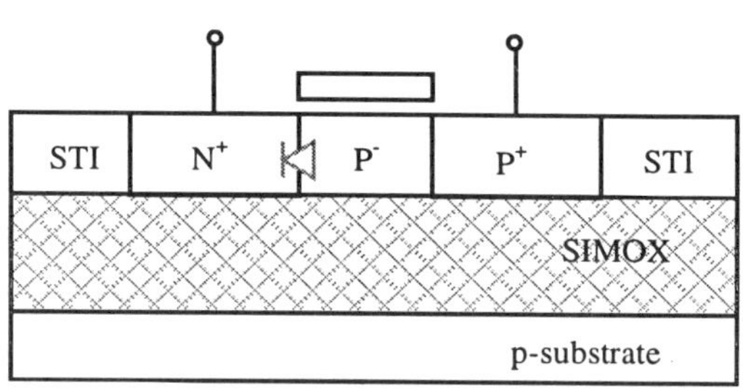

Figure 7.43 A diode ESD protection structure for SOI applications [28] (*Reproduced here by kind permission of Elsevier Science and authors*).

In SiGe technology, the main advantage of using SiGe comes from the fact that it features narrower band-gap and much higher mobility. A novel ESD protection structure using this SiGe property is illustrated in Figure 7.44, where a thin SiGe layer is place at the bottom of the NMOSFET channel between the source and drain. Because of its narrow band-gap, the current will be confined to the SiGe layer and flows deep in the silicon. Since the heat generation is pushed downward into the bulk silicon where heat can be dissipated out of the bottom easily, improvement in ESD protection can be realized. Study shows that the parasitic lateral NPN bipolar transistor can be turned on faster and its current gain is 10 times that of a regular NMOS ESD protection device [30].

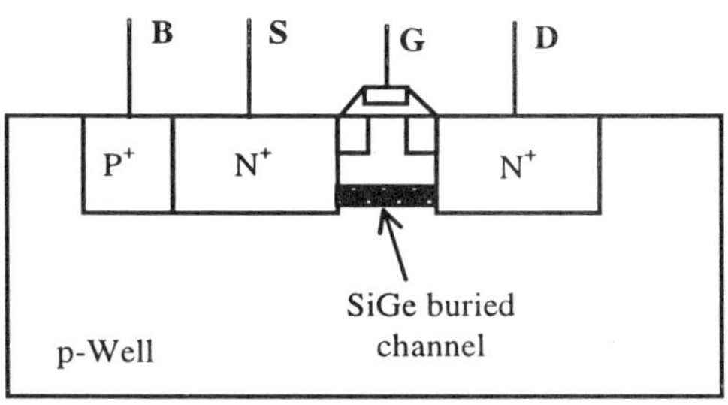

Figure 7.44 A deep SiGe layer is placed at the bottom of the channel of a NMOSFET ESD protection structure that pushes the heat source deep into the bulk, resulting in better ESD protection performance [30] (*Reproduced here by kind permission of IEEE and authors*).

7.8. ESD PROTECTION FOR NANO TECHNOLOGY

ESD protection design for VDSM IC technologies at sub-180nm nodes is very challenging considering the micro-heat-spreading effect that becomes dominating in ultra small feature size devices. For example, study shows that the heat distribution from a hot spot at the drain junction may well reach out into the source contact area and causes substantial local temperature increase, resulting ESD damage in metal contacts [31]. The golden rule of using a minimum SCGS spacing must be re-considered in the VDSM regime. ESD phenomena and protections for future nano-scale technologies is even tougher a design challenge ahead. Other than the problems facing VDSM IC technologies, extra challenging issues emerge at nano-scale (<50nm node). One of them is associated with phonon-enhanced micro-heat spreading effect. In nano-regime, device characteristic dimensions and hotspot sizes become comparable to the phonon mean-free-path or wavelength, the ballistic phonon transport/scattering plays a much larger role in overall electro-thermal conduction, currently dominated by the charge transport theory [32]. The thermal-damage-based ESD failure becomes more severe that may completely break up all the existing IC ESD design rules. Investigation of the phonon-enhanced micro-heat spreading effects is currently drawing research attentions. Novel protection concepts may be expected for nano-scale ESD protection in near future.

7.9. SUMMARY

In summary, both layout design and technology evolution will strongly influence ESD protection circuit design and performance. In terms of layout, two basic considerations should be kept in mind when doing ESD protection circuit design, i.e., to ensure uniform current distribution and to make area-efficient layout. Practical ESD-enhancement layout techniques include to use smoothed geometries, to set up adequate critical spacing, to properly place contacts and vias, as well as to optimise metal interconnect routing, and so on. Use of bounding pad oriented ESD protection is a very attractive solution to high-pin-count area-sensitive chips. In regarding technological impacts, every single new process techniques designed to boost transistor operation should be evaluated against its influences on ESD protection performance. Most advanced process techniques, such as, LDD and silicidation, can degrade ESD robustness dramatically. Technology scaling generally makes ESD protection circuit design more difficult. Many after-qualification fixing measures are available to recover the lose in ESD protection performance, such as LDD-blocking, silicide-blocking and ESD implant, etc, at extra costs. It is wise to consider technology-ESD co-development in early phase of new process development, where it is possible to strike a balance between regular transistor operation and ESD robustness. In near future, ESD protection for VDSM and nano-scale technologies is a challenge to be addressed.

REFERENCES

[1] Wang, A. Z., Feng, H. G., Gong, K., Zhan, R. Y. and Stine, J., "On-Chip ESD Protection Design for Integrated Circuits: an Overview for IC Designers", *J. Microelectronics, Elsevier Science, Vol. 32/9,* August 2001, pp.733-747.

[2] Amerasekera, A. and Duvvury, C., *ESD in Silicon Integrated Circuits*, John Wiley & Sons, England, 1995.

[3] Dabral, S. and Maloney, T., *Basic ESD and I/O Design,* John Wiley & Sons, 1998.

[4] Feng, H. G., "A Mixed-Mode Simulation-Design Methodology for on-Chip ESD Protection Design", *MS Thesis*, Illinois Institute of Technology, May 2001.

[5] Rountree, R. and Hutchins, C., "NMOS Protection Circuitry", *IEEE Trans. Elec. Dev., ED-32,* 1985, pp.910-917.

[6] Palella, A. and Domingos, H., "A Design Methodology for ESD Protection Networks", *Proc. EOS/ESD Symp.,* 1985, pp.169-174.

[7] McPhee, R., Duvvury, C., Rountree, R. and Domingos, H., "Thick Oxide ESD Performance under Process Variations", *Proc. EOS/ESD Symp.,* 1986, pp.173-179.

[8] Wilson, D., Domingos, H. and Hassan, M., "Electrical Overstress in nMOS Silicided Devices", *Proc. EOS/ESD Symp.,* 1987, pp.265-273.

[9] May, J. and Guravage, J., "Interpretation of EPS Induced Discoloration in ICs", *Proc. Int'l Symp. Testing Fail. Anal.,* 1990, pp.143-147.

[10] Wang, A., Tsay, C. and Deane, P., "A Study of NMOS Behaviors under ESD Stress: Simulation and Characterization", *Microelectronics Reliability, 38, Elsevier* Science, pp.1183-1186, 1998.

[11] Duvvury, C., Rountree, R and McPhee, R., "ESD Protection: Design and Layout Issues for VLSI Circuits", *IEEE Tran. Industry Appl., Vol. 25, No. 1*, January 1989, pp. 41-47.

[12] Ker, M., Wang, K, Joe, M., Chu, Y. and Wu, T, "Area-Efficient CMOS Output Buffer with Enhanced High ESD Reliability for Deep Submicron CMOS ASIC", *Proc. 8th Annual IEEE Int'l ASIC Conf.,* 1995, pp. 123-126.

[13] Baker, L. Currence, R., Law, S., Le, M., Lee, C., Lin, S. and Teene, M., "A Waffle Layout Technique Strengthens the ESD Hardness of the NMOS Output Transistor", *Proc. EOS/ESD Symp.,* 1989, pp.175-181.

[14] Ker, M., Wu, C. and Wu, T., "Area-Efficient Layout Design for CMOS Output Transistors", *IEEE Trans. Elec. Dev., Vol. 44, No. 4*, April 1997, pp.635-645.

[15] Wang, A. Z. and Tsay, C. H., "A Compact Square-Cell *ESD* Structure for BiCMOS ICs", *Proc. IEEE BCTM*, 1999, pp. 46-49.

[16] Wang, A., Tsay, C., Bielawski, J. and DeClue, L., "Design Optimization of a Practical ESD Protection Circuit by CAD: A Case Study", *Proc. IEEE Univ. Gov. Indust. Microelec. Symp.*, 1999, pp. 116-119.

[17] Anderson, W., Gonzalez, M., Knecht, S. and Fowler, W., "ESD Protection under Wire Bonding Pads", *Proc. EOS/ESD Symp.,* 1999, pp.88-94.

[18] Feng, H., Zhan, R., Gong, K. and Wang, A. Z., "A New Pad-Oriented Multiple-Mode ESD Protection Structure and Layout Optimisation", *IEEE Electron Device Letters, Vol. 22, No. 10,* October 2001, pp.493-495.

[19] Feng, H., Zhan, R., Gong, K. and Wang, A., "A Pad-Based Low-Parasitic Multiple-Mode ESD Protection Structure for ICs", submitted to *IEEE Trans. CAS II*, 2001.

[20] Voldman, S. and Gross, V., "Scaling, Optimization and Design Considerations of Electrostatic Discharge Protection Circuits in CMOS Technology", *Proc. EOS/ESD Symp.,* 1993, pp.251-260.

[21] Amerasekera, A. and Duvvury, C., "The Impact of Technology Scaling on ESD Robustness and Protection Circuit Design", *Proc. EOS/ESD Symp.,* 1994, pp.237-245.

[22] Shabde, S., Simmons, G., Baluni, A. and Back, D., "Snapback Induced Gate Dielectric Breakdown in Graded Junction MOS Structures", *Proc. Int'l Relia. Phys. Symp.,* 1984, pp.165-168.

[23] Daniel, S. and Krieger, G., "Process and Design Optimization for Advanced CMOS I/I ESD Protection Devices", *Proc. EOS/ESD Symp.,* 1990, pp. 206-213.

[24] Amerasekera, A., McNeil, V. and Rodder, M., "Correlating Drain Junction Scaling, Salicide Thickness, and Lateral NPN Behaviour, with the ESD/EOS Performance of a 0.25μm CMOS Process", *Digest IEDM,* 1996, pp.893-896.

[25] Okushima, M., Noguchi, K., Sawahata, K., Suzuki, H., Kuroki, S., Koyama, S., Ando, K. and Ikezawa, N., "ESD Protection Scheme using CMOS Compatible Vertical Bipolar Transistor for 130nm CMOS Generation", *Digest IEDM,* 2000, pp.127-130.

[26] Gong, K. Feng, H. G., Zhan, R. Y. and Wang, A. Z., "A Study of Parasitic Effects of ESD Protection on RF ICs", in press, *IEEE Trans. Microwave Theory and Techniques,* Feb. 2002.

[27] Chan, M., Yuen, S., Ma, Z., Hui, K., Ko, P. and Hu, C., "Comparison of ESD Protection Capability of SOI and Bulk CMOS Output Buffers", *Proc. Int'l Relia. Phys. Symp.,* 1994, pp. 292-298.

[28] Voldman, S., Assaderaghi, F., Mandelman, J., Hsu, L. and Shahidi, G., "Dynamic Threshold Body- and Gate-Coupled SOI ESD Protection Networks", *J. Electrostatics, Elsevier Science, 44,* 1998, pp. 239-255.

[29] Verhaege, K., Groesenken, G., Colinge, J. and Maes, H., "Double Snapback in SOI nMOSFET's and its Application for SOI ESD Protection", *IEEE Elec. Dev. Lett., Vol. 14, No. 7,* July 1993, pp. 326-328.

[30] Choi, C., Park, Y., Lee, S. and Kim, K., "Novel ESD Protection Transistor Including SiGe Buried Layer to Reduce Local Temperature Overheating", *IEEE Trans. Elec. Dev., Vol. 43, No. 3,* March 1996, pp. 479-489.

[31] Feng, H. G., Gong, K. and Wang, A., "ESD Protection Design Using Copper Interconnects: More Robustness and Less Parasitics", *SRC Publication: 2000 Publications in Copper Design Challenge, Pub. P000375,* 17-Mar-2000.

[32] Chen, G., "Non-local and Nonequilibrium Heat Conduction in the Vicinity of Nanoparticles", *J. Heat Transfer, Vol. 118,* pp.539-545, 1996.

Chapter 8

ESD SIMULATION-DESIGN METHODOLOGIES

Successful ESD protection circuit design depends upon the design approaches used. Various ESD protection circuit design methods are available, ranging from traditional trial-&-error approach, to rational, however, limited CAD-based ESD simulation method, to more accurate mixed-mode ESD simulation methodology. These ESD design approaches have different level of implementation complexity, require different degree of design experiences and knowledge, and deliver different level of design success. This chapter discusses, in depth, these various ESD protection design methodologies. Practical ESD protection design examples are provided.

8.1. ESD PROTECTION DESIGN METHODS: TRIAL-&-ERROR *versus* PREDICTIVE

On-chip ESD protection circuit design is a fairly complex design task because of the fact that ESD phenomena involve multiple level coupling effects, i.e., process-device-circuit-electro-thermal couplings. This multiple level coupling nature means that all these contributing factors, i.e., process, device, circuit, electro and thermal properties, must be dealt with simultaneously in an integrated fashion. Unfortunately, ESD protection circuit design is not as simple as it sounds like. Lack of proper CAD simulation tools, sufficient computing power and special ESD protection device models prevents circuit designers from performing meaningful and practical simulation during ESD protection circuit design. Currently, the experience-based trial-and-error approaches still dominate the ESD protection circuit design practices. While experience certainly plays a useful and significantly role, solely experience-based design approach does not guarantee design successes, particularly the highly desired first-time design success. A traditional trial-and-error design approach is illustrated in Figure 8.1, where the design procedures start with defining the design specifications. With the aid of existing ESD protection design experiences, an IC designer begins design of the ESD protection circuits. Some limited and isolated ESD simulation may be performed during the design, i.e., either device level simulation or circuit level simulation. The design will be taped-out afterward and silicon wafers will be started. Testing and debugging work will be performed after receiving the silicon wafers. Should one be lucky enough, the design may work the first time. Otherwise, one has to fix the bugs and repeat the design procedures until it functions. The main problem behind the trial-and-error design approach is that, as mentioned in previous chapters, ESD protection circuit design is non-portable in nature. A proven successful ESD protection circuit design for a specific product using a specific process technology usually does not ensure functionality of an ESD protection design for a different IC chip even using the same process technology. Informal statistics indicate that it takes average three silicon iterations for an experienced IC designer to complete a successful ESD protection circuit design [1]. The main disadvantage of the trial-and-error design approach is

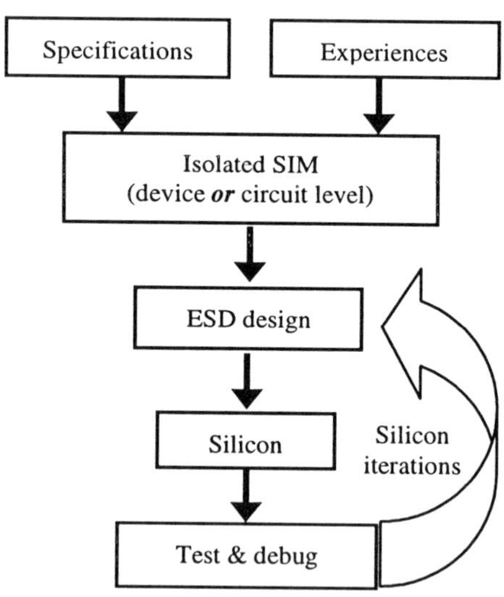

Figure 8.1 A design flow chart for the traditional trial-and-error ESD protection circuit design approach.

that it is very time-consuming and extremely costly. In fact, ESD protection design becomes a main dragging-foot factor in modern IC product development. It is imperative to develop a more efficient, predictive and practical ESD protection design methodology that can, well, ideally, guarantee the first time design success. ESD design simulation is a critical factor in achieving the goal of predictive ESD protection circuit design. Several ESD protection design and simulation approaches are proposed that provide ESD protection design prediction to various degrees [2-8]. These methods are discussed in the following sections.

8.2. ESD DESIGN-SIMULATION: DEVICE LEVEL *versus* CIRCUIT LEVEL

A rational ESD protection circuit design approach requires comprehensive ESD simulation in order to realize ESD protection circuit design prediction and to achieve first-time design success. There are two types of ESD simulation approaches existing currently: device level ESD simulation and circuit level ESD simulation. The need for device level ESD simulation is fairly straightforward. Since ESD phenomena involve electro-thermal coupling effects that are governed by semiconductor device physics equations [9], i.e., Poisson's equation (8.1), Boltzman transport equation (8.2), continuity equation (8.3) and lattice temperature heat flow equation (8.4) as listed below,

$$\varepsilon\nabla^2\Psi = -q(p-n+N_D^+-N_A^-)-\rho_s, \tag{8.1}$$

$$\vec{J}_{(n,p)} = -q\mu_{(n,p)}(n,p)\vec{\nabla}\phi_{(n,p)}, \tag{8.2}$$

$$\frac{\partial(n,p)}{\partial t} = \frac{\pm 1}{q}\vec{\nabla}\cdot J_{(n,p)} - U_{(n,p)} = F_{(n,p)}(\Psi,n,p), \tag{8.3}$$

and

$$\rho c\frac{\partial T}{\partial t} = H + \vec{\nabla}\{\lambda(T)\vec{\nabla}T\}, \tag{8.4}$$

where Ψ is potential; n & p are electron and hole carrier densities; N_D^+, N_A^-, ρ_s are ionised impurity densities and surface charge density; J, U, H, ρ, c & μ are current density, carrier generation/recombination rate, heat source, material mass, specific heat and mobility, respectively.

The way device level ESD simulation works is that a simulator is used to generate a mesh across the ESD protection device and then to solve the above semiconductor device physics equations simultaneously across the mesh generated under an ESD stress. Such device simulation technique is referred to as technology CAD (TCAD) or numerical CAD [10]. The semiconductor lattice node temperatures across the mesh are extracted and examined against the pre-set ESD failure criterion, commonly being the silicon melting temperature of 1421 °C. If the maximum lattice temperature at any node exceeds the critical temperature, the ESD protection device fails. Several device level ESD simulation approaches have been reported [2-4] that use DC and/or square pulses to stress the ESD protection structures as illustrated in Figure 8.2. While such device level ESD simulation is essential to understanding ESD phenomena, its main disadvantages are that it isolates the ESD protection device from the circuit and uses DC or square-wave stresses to emulate real-world ESD events. This approach is usually used to study ESD operation after silicon, a backward investigation, instead of to predict ESD protection circuit design in a forward design manner.

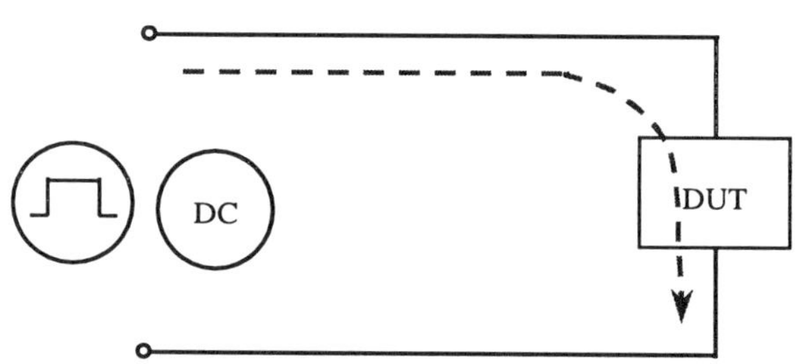

Figure 8.2 In device level ESD simulation, an ESD protection structure is treated as a stand-alone device, which is stressed by DC and/or square-wave sources to emulate an ESD event.

Circuit level ESD simulation is preferred by ordinary IC designers because of the following reasons: circuit simulation is less computing-intensive and more efficient, and, well, not every circuit designer can understand the complex device physics behind the TCAD results. Normal circuit simulation techniques, such as, SPICE, have been used for ESD design. While SPICE simulation may be useful in ESD design for things like timing and selection of resistance and capacitance values, for example, in the gcNMOS ESD protection structure, one cannot perform complete ESD simulation using SPICE because of its deficiency in dealing with ESD-critical features, such as avalanche snapback and thermal re-distribution. In principle, circuit level ESD simulation is similar to regular SPICE-based circuit simulation, where a device is treated as a black-box element with its terminal electronic characteristics being described by a device model file and the Kirchhoff loop-voltage and node-current equations are solved by a mathematical solver, e.g., a SPICE simulator, under given biasing conditions. These circuit equations are given by the Kirchhoff voltage and current laws as,

$$\sum_{x=1}^{m} i_x = 0\,, \quad m = \text{number of paths at one node,} \tag{8.5}$$

and

$$\sum_{y=1}^{n} v_y = 0\,, \quad n = \text{number of branches in a closed loop.} \tag{8.6}$$

Ideally, as long as accurate device models are available for the ESD protection devices, and thermal distribution equation is included in the equation set, ESD circuit level simulation is, mathematically, nothing different from that of regular circuit simulation. Several unique and critical concerns associated with ESD circuit simulation must be addressed. Accurate ESD device models are essential to successful ESD circuit simulation. However, ESD device modelling is still a subject under research. The difficulties in ESD device modelling include high-current behaviours, avalanche snapback, heat generation and distribution, as well as failure criteria, etc, which are striking different from regular device models, such as the popular SPICE device models. The ESD device modelling problem will be discussed in the next section. In addition, it may not be practically feasible to define and model the heat sources, or, the hot spots, associated with ESD stresses within an ESD protection structure beforehand, which is required to solve the thermal distribution equation. A typical heat source model is depicted in Figure 8.3 for a traditional MOSFET ESD protection device, where heat generation under ESD stresses is believed to be located at the drain junction edge facing the conducting channel and is modelled by an

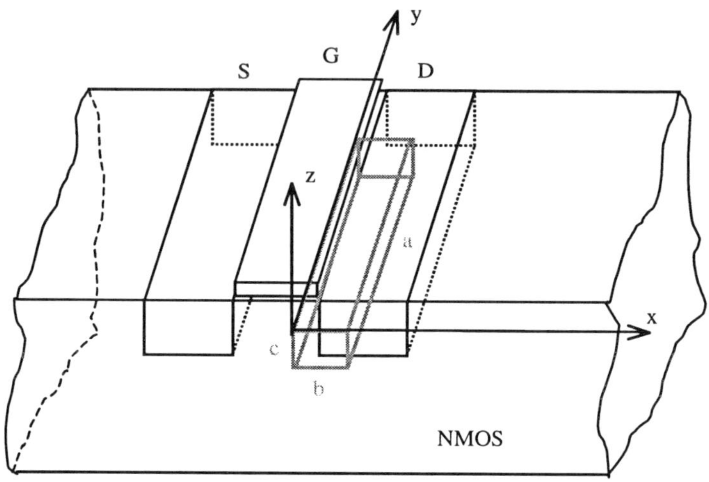

Figure 8.3 A simplified heat source model for MOSFET device uses a parallelepiped heat source to model heat generation under ESD stresses.

over-simplified parallelepiped with physical dimensions of a, b and c. The thermal equation is then solved by the simulator for device temperatures, which can be used as the ESD failure criterion. Such simulation approach has been implemented in ESD circuit simulators, such as iETSIM [7, 11, 12]. The challenges in circuit level ESD simulation are to develop accurate ESD device models and to precisely define the heat sources in practical, often unknown, ESD protection structures beforehand, if ever possible.

8.3. ESD PROTECTION DEVICE MODELING

Circuit level electro-thermal simulation for ESD protection circuit may be possible provided accurate ESD device models are available for the ESD protection structures used. ESD device models should address unique ESD operation properties, such as, high current, high voltage, electro-thermal coupling, avalanche snapback, etc, which are not included in regular SPICE device models. Typically, avalanche multiplication model can be used to account for snapback behaviour; lumped thermal element network can be used to describe thermal behaviours with thermal resistance and capacitance being defined according to geometry, boundaries and thermal properties. Electro-thermal coupling can be established based upon a current source to power dissipation to heat generation correlation. In a typical example, a MOSFET ESD protection device can be modelled by a lateral bipolar junction transistor (BJT) during ESD stresses [11, 13-15]. However, major disadvantage exists in these ESD device models because they do not address the critical geometry-sensitivity issue in ESD protection structures. Because of the highly localized heat generation feature due to extremely short ESD pulse duration, ESD damages often occur at the edges and/or corners of devices. This geometry-sensitive ESD failure phenomenon can not be addressed by existing ESD device models. To address the critical geometry-sensitivity problem of ESD protection structure, a parallel section-wise device model is introduced to account for ESD damages at edges and corners of ESD protection structures. Taking a MOSFET ESD protection device shown in Figure 8.4 as an example, instead of using a uniform lateral BJT transistor as the operating device, a group of parallel BJT transistors may be used to model the MOSFET transistor. The MOSFET transistor is divided into several sections with individual BJT devices, connected in parallel, representing central, edge and corner sections of the MOSFET transistor. In ESD operation, these parallel section-wise BJT devices operate independently and the localized ESD damage effect can be modelled by the

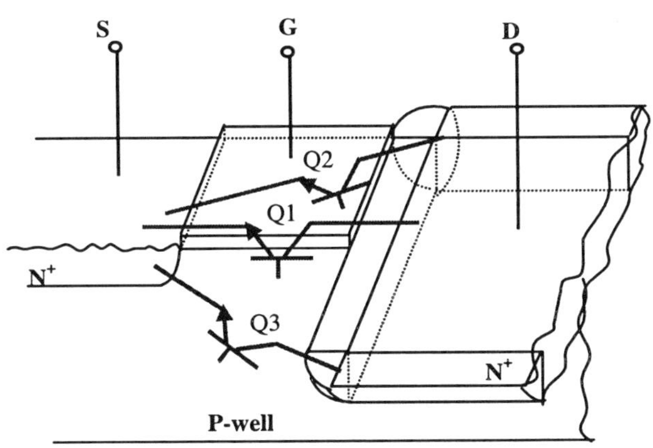

Figure 8.4 A parallel section-wise model for a MOSFET ESD protection structure uses a parallel BJT network to account for the geometry-sensitivity ESD damage effect.

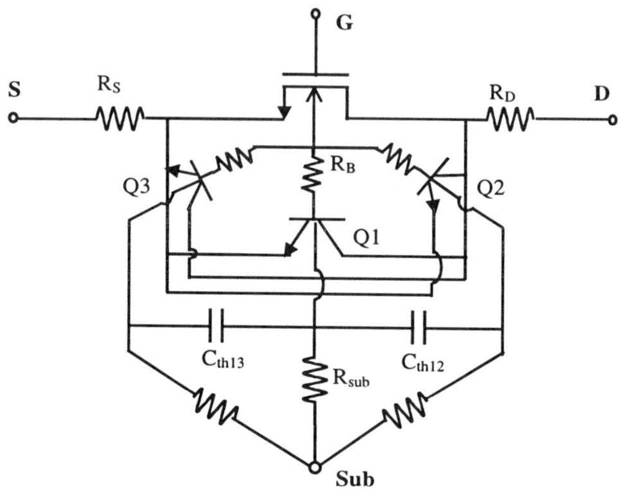

Figure 8.5 An equivalent circuit for the parallel section-wise model for MOSFET.

competing behaviours of these parallel BJT devices to take into consideration of the edge-corner ESD failure phenomena. An equivalent circuit for the parallel section-wise model is given in Figure 8.5, where thermal capacitors, C_{th12} and C_{th13}, are introduced to depicted the thermal coupling effect between the local heat sources in different sections. A SPICE-compatible model for each BJT is described in Figure 8.6, where, in addition to the regular forward and reverse currents I_F and I_R, new current sources, I_{avC} and I_{avE}, account for avalanche breakdown at either collector or emitter junction of the BJT during positive or negative ESD stressing; while current sources, I_{thC} and I_{thE}, reflect thermal generation of power dissipation due to avalanche multiplication. A symmetric model schematic is used to account for both positive and negative ESD pulses. Since heating mainly comes from power dissipation of avalanche currents and heat dissipation through both silicon bulk substrate and the surface underneath gate dielectric film, ESD thermal phenomena may be modelled by a thermal circuit shown in Figure 8.7, where electro-thermal coupling can be depicted by equation (8.7), where T is lattice temperature, t is time, T_{amb} is ambient temperature, P is power dissipation, R_{th} and C_{th} are thermal resistance and capacitance, respectively. To ensure simulation continuity, the Miller formula is modified as equation (8.8) [16], where a_1, a_2, b_1 and b_2 are fitting parameters.

$$T = T_{amb} + PR_{th} - PC_{th}\frac{dT}{dt} \tag{8.7}$$

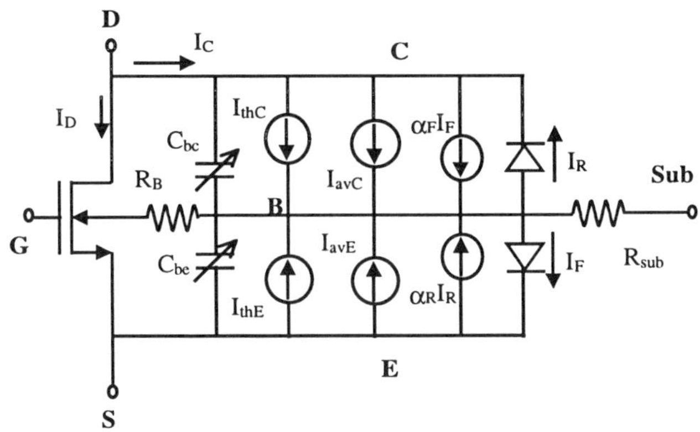

Figure 8.6 A SPICE-compatible model for each parasitic BJT transistor in the MOSFET parallel section-wise model circuit consists of regular currents, I_F and I_R, and additional avalanche breakdown currents, I_{avC} and I_{avE}, as well as thermal current sources, I_{thC} and I_{thE}.

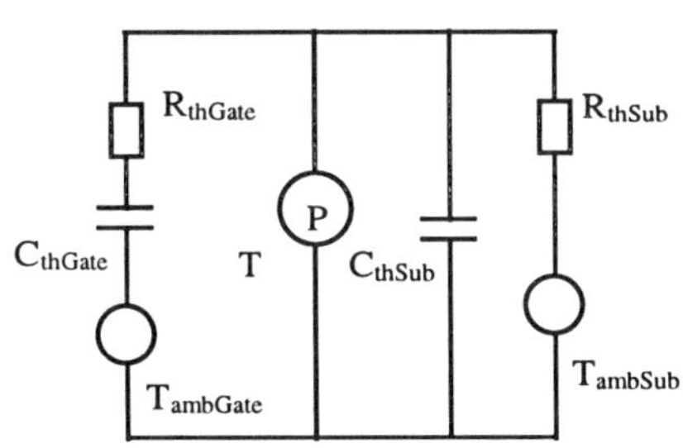

Figure 8.7 A thermal circuit model for MOSFET ESD protection device accounts for heat dissipation through both bulk substrate (with footnote "Sub") and subsurface underneath the gate (with footnote "Gate").

$$M = e^{\{a_1 (V_D - V_{Dsat} - b_1)\}} + e^{\{a_2 (V_D - V_{Dsat} - b_2)\}}. \qquad (8.8)$$

Time-dependent latent dielectric damage is another major ESD failure problem of particular importance to advanced CMOS technologies. In one latent ESD failure model derived from the charge trapping theory [17], the total amount of charge defects accumulated within dielectric films are evaluated and used as an indicator for gate oxide damage. Use the 1/E model, where E is electric field, typical latent defect parameters such as the time-to-ESD damage, t_{BD}, the latent failure probability, F, and the associated time-varying failure can be described by,

$$t_{BD} = \kappa e^{\frac{\delta}{E_{ox}}} = \kappa e^{\frac{\delta H_{ox} - \Delta H_{ox}}{V_{ox}}}, \qquad (8.9)$$

$$F = 1 - e^{-AN_{ESD}(\Delta H_{ox})} e^{-AN(\Delta H_{ox})}, \qquad (8.10)$$

and

$$1 = \frac{1}{\kappa} \int_0^{t_{BD}} e^{\frac{\delta H_{ox} - \Delta H_{ox}}{V_{ox}}} dt + \frac{Q_{ESD}}{t}, \qquad (8.11)$$

where H_{ox} and ΔH_{ox} are gate oxide thickness and its defect-thinning parameter, N is intrinsic defect density, N_{ESD} is ESD-induced defects, and Q_{ESD} is ESD stressing induced defects. The thermal effect is described by coefficients κ and δ, which are given by,

$$\kappa(T) = \kappa_o e^{-\frac{E_a}{k}\left(\frac{1}{T} - \frac{1}{300}\right)}, \qquad (8.12)$$

and

$$\delta(T)=\delta_o\left\{1+\gamma\left(\frac{1}{T}-\frac{1}{300}\right)\right\}, \tag{8.13}$$

where E_a is activation energy and γ is a constant. All fitting factors can be extracted experimentally. An ESD latent failure rate monitor, F(~t_{BD}), can be modelled by Figure 8.8. Alternatively, a defect-related resistance, R_{BD} (Q_{BD}), calculated as a post-resistance associated with latent breakdown defects, Q_{BD}, may be used as a latent defect monitor, since the post-resistance is related to the charge-to-damage due to ESD stresses [18].

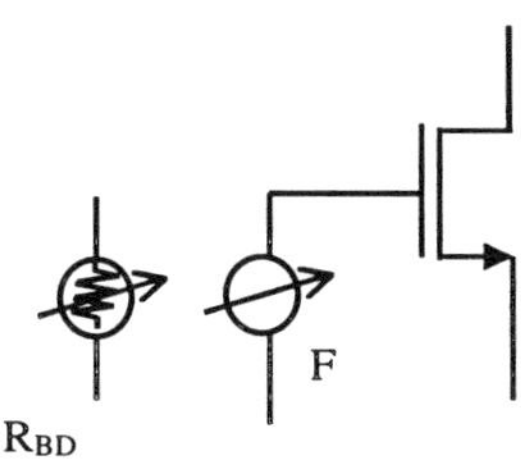

Figure 8.8 A latent dielectric ESD failure model uses a latent failure rate monitor or a post-resistor monitor to depict ESD-induced latent defects in gate oxide layers.

Ideally, using the parallel model and latent defect model described previously, one should be able to investigate the geometry-sensitive ESD failure problem and latent ESD damage phenomena in simulation. "*Now, wait a minute, tell me where do I get those nice models?*" Good question! Unfortunately, such ESD device models are still under research.

8.4. MIXED-MODE ESD SIMULATION FOR DESIGN PREDICTION

Considering the multiple level coupling nature of ESD protection operation, it is imperative to perform ESD simulation in a mixed mode, which can deal with the process-device-circuit-electro-thermal coupling issues altogether and simultaneously to ensure design successes, because all these factors play significant roles in ESD protection operation [19-21]. Typical mixed-mode ESD simulation procedures are depicted by the diagram shown in Figure 8.9, where initial inputs are the technology and design specifications, as well as experiences. However, the main design workload is to perform comprehensive ESD simulation at both device and circuit levels that also include electro-thermal behaviours and process properties. To ensure design prediction, full-scale calibration must be conducted during ESD simulation. One would not start real ESD protection circuit design and any silicon wafers until the design works on the screen. This mixed-mode ESD simulation approach is the way leading to first-time design success.

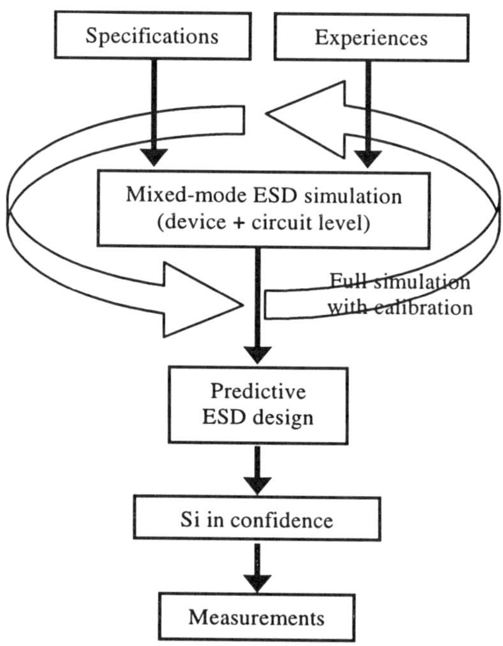

Figure 8.9 A typical mixed-mode ESD simulation flow diagram.

Several such mixed-mode ESD simulation-design methodologies, using advanced TCAD software with circuit simulation capabilities [22-25], are available for practical on-chip ESD protection circuit design [20, 21, 26]. The mixed-mode ESD design-simulation methodology involves multiple-level coupling in simulation, i.e., combined process-device-circuit-electro-thermal considerations. Mixed-mode ESD simulation is so entitled because it involves process simulation, device characteristic simulation and circuit performance simulation in a closed-loop fashion. The idea behind is to create actual ESD protection structures by process simulation, evaluate individual ESD structure by device simulation, and simulate ESD circuit performance under real-world transient ESD stresses by mixed device-circuit level simulation. From the design viewpoint, one ought to define the critical I-V characteristic parameters, such as, triggering point (V_{t1}, I_{t1}), triggering timing (t_1), holding point (V_h, I_h), and failure threshold (V_{t2}, I_{t2}), etc. The goal is to predict ESD performance in pre-silicon design phase. For ESD protection circuit design prediction, calibration is a critical factor in ESD simulation, same as in typical SPICE simulation cases. The simulation calibration work must be conducted carefully for every new design, either using different technologies or the same technology. The reason, often overlooked by designers, is that ESD protection circuit design is actually custom design in the sense that it is always sensitive to structures, layout, circuits, technologies, and applications. The key point is that ESD protection solution is not portable. A complete mixed-mode ESD simulation-design flow chart is illustrated in Figure 8.10. The ESD design flow starts with input of specifications for the processes, IC circuits and ESD requirements. Process specifications provide designers with technology information such as, what technology is used (e.g., bipolar, CMOS, BiCMOS, etc), and what diffusion layers are available for ESD protection design including doping profiles and diffusion thickness, etc. These process parameters set up the boundaries where designers can play around. Circuit specifications tell designers what protection scheme may be used and what protection structure cannot be used. For example, a regular ggNMOS ESD protection device may not be used for high-voltage ICs because of the short-circuit problem. A protection structure with large parasitic capacitance is not suitable for RF IC chips. ESD protection specifications define critical parameters such as trigger voltage, response time, leakage current, ESD protection level, etc. Next, one starts initial selection of possible ESD protection structures, followed by initial ESD simulation and ESD protection structure test pattern design. The initial selection of possible ESD protection structures is usually unrealistically large. However, most of them will be eliminated during initial simulation, ending up with a few most likely ESD protection structure types to work with. The ESD test patterns should be as simple as possible, typically being single-finger small structures, which are used for full-scale ESD simulation calibration. Simulation calibration serves to verify ESD simulation programs including critical device coefficient tune-up. The calibrated simulation codes are then used to perform the real ESD protection

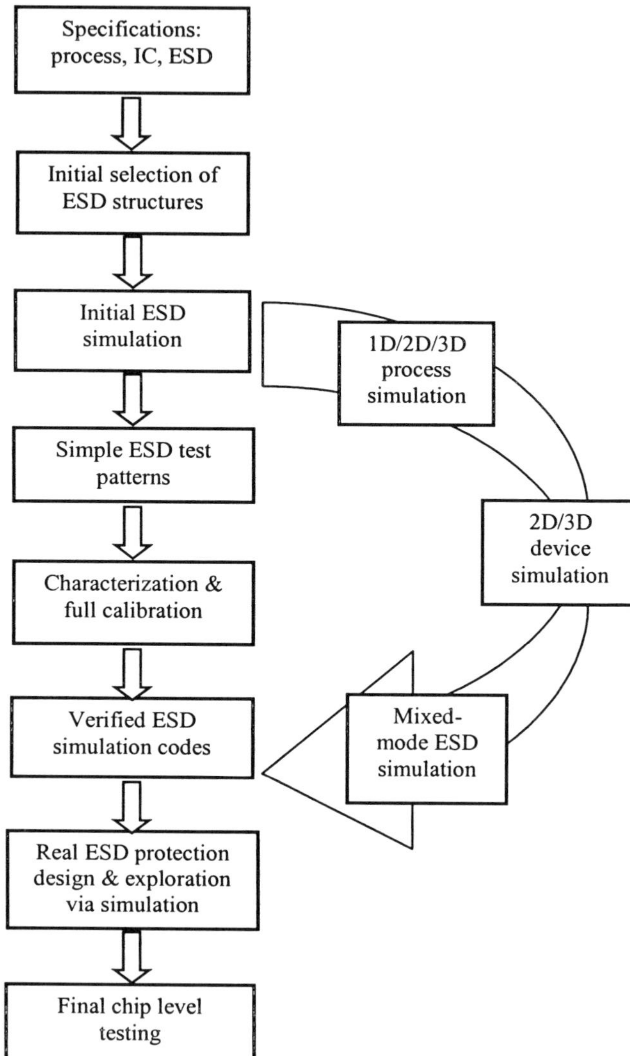

Figure 8.10 A flow chart for mixed-mode ESD simulation.

circuit design. With functionalities confirmed by ESD simulation, one starts to run silicon wafers and finally, conduct ESD measurements. An ESD failure criterion needs to be defined for ESD simulation. One commonly used ESD failure criterion, though still arguable, is the melting temperature of the material involved. For example, melting temperature of 1685 °K for silicon, eutectic temperature of 823 °K for aluminium and 1357 °K for copper metal interconnect. With the melting temperature as ESD failure criterion, ESD simulation will extract lattice temperatures across the protection structure and reports a failure when a maximum temperature exceeds the critical temperature anywhere within the protection structure. Of course, latent ESD failure in dielectric films cannot be simulated using the critical temperature failure criterion. With the aid of comprehensive ESD simulation at process, device and circuit levels, reasonable ESD protection circuit design prediction may be realized to achieve the first-time design success. One unique feature of the mixed-mode ESD simulation is that it uses given ESD model circuits (i.e., HBM, MM, CDM and IEC model circuits as described in Chapter 2) to generate ESD pulses that are used to stress the ESD protection circuit during simulation without pre-assuming a heat source, as is the case in Section 8.2. This non-assumption approach makes the mixed-mode ESD simulation approach be more close to the reality, therefore makes the ESD design prediction be more feasible. Figure 8.11 illustrates a

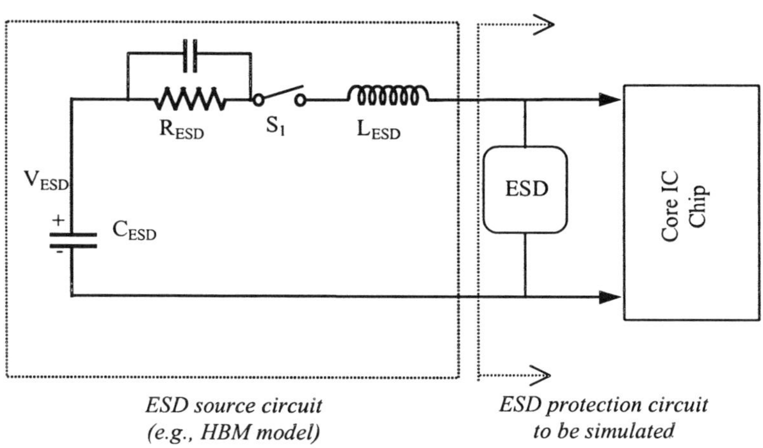

Figure 8.11 A typical schematic for mixed-mode ESD simulation uses an HBM ESD model circuit to generate ESD pulse that is used to stress the ESD protection structure.

schematic for the mixed-mode ESD simulation approach, where an HBM ESD model circuit is used. The mixed-mode ESD simulation includes both static simulation and transient simulation. Static ESD simulation is usually performed first to quickly study key parameters of an ESD protection structure, such as, trigger point, snapback behaviour and holding point, etc. Static ESD simulation results may be matched with curve tracing measurement data in calibration. Transient ESD simulation is essential because ESD event is transient in nature, which can only be simulated in transient mode. Transient TLP measurement data are used to compare with transient ESD simulation results during simulation calibration. The simulated thermal damage threshold will also be calibrated with the final ESD zapping results. Figure 8.12 shows a complete testing diagram that uses DC curve tracer, transient TLP tester and ESD zapping tester. Practical ESD design examples using mixed-mode ESD simulation approach are given in the following section.

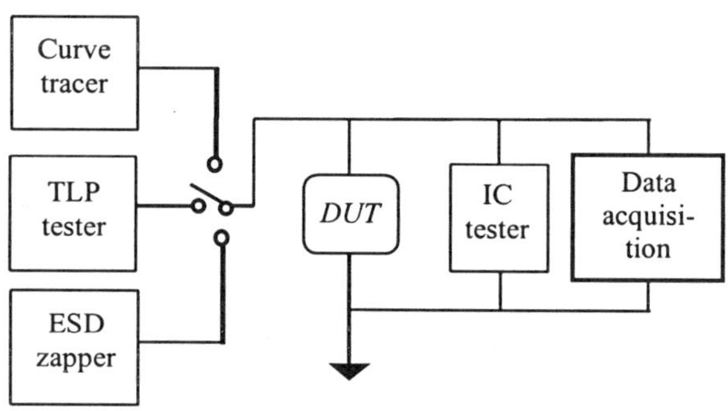

Figure 8.12 A complete ESD testing diagram includes DC measurement using a curve tracer, transient testing using a TLP tester and stress testing using an ESD zapping tester.

8.5. MIXED-MODE ESD SIMULATION: CASE STUDY

To illustrate how to efficiently use the mixed-mode ESD simulation methodology in practical ESD protection circuit design, A few real-world ESD protection design examples are discussed in this section.

8.5.1. Understanding ESD Simulation Results

Let us first understand the general flow of the mixed-mode ESD simulation and interpret the simulation results. As an example, a ggNMOS ESD protection structure designed and implemented in a commercial six-metal 0.18μm CMOS process is presented [27]. Figure 8.13 shows a cross-section of the ggNMOS ESD protection structure obtained from process simulation, where a small source-contact-to-gate spacing, SCGS=0.36μm, and a optimised large drain-contact-to-gate spacing, DCGS=2.15μm, are selected according to the discussion in Chapter 7. These optimal values are obtained by conducting ESD simulation, not directly from experiences. It is worth noting that the SCGS value is not the minimum values as the traditional rule-of-thumb suggested, because the thermal spreading effect may cause over heating in the source contact region in sub-quarter-micron technologies [21, 27]. Device level simulations are conducted next under DC

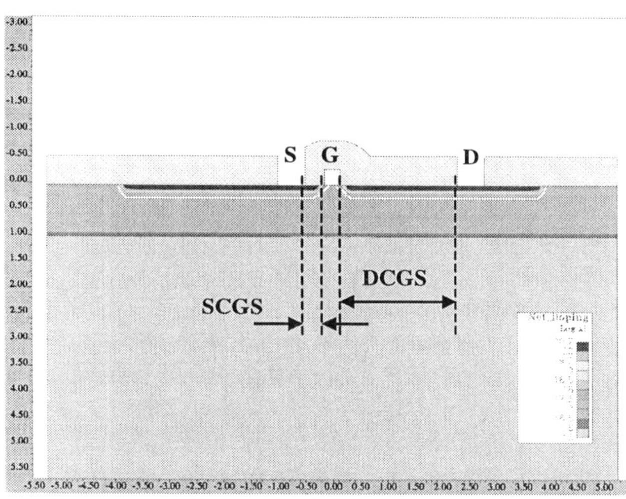

Figure 8.13 A cross-section of a ggNMOS ESD protection from process simulation shows a small SCGS spacing and a large DCGS spacing [27].

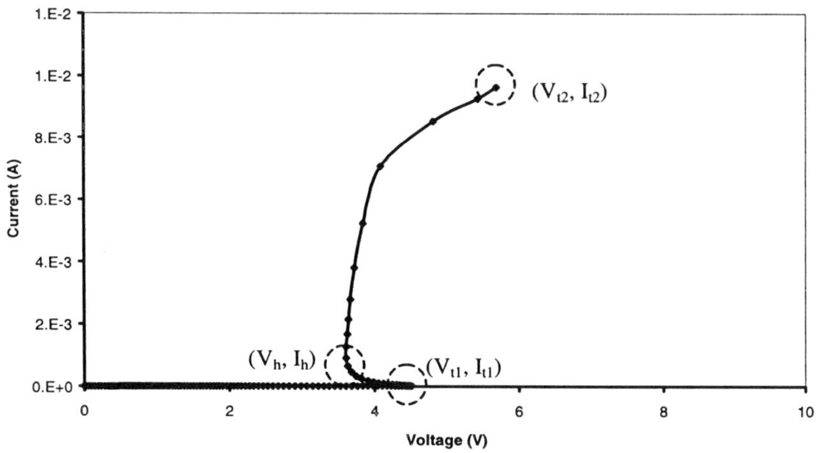

Figure 8.14 I-V characteristic from DC sweeping shows critical parameters, e.g., (V_h, I_h), and (V_{t2}, I_{t2}), etc. Thermal breakdown is not clear in this example due to simulation resolution.

stresses for a general picture of its I-V characteristics. Figure 8.14 is a typical I-V characteristic by DC sweeping, where critical parameters, e.g., trigger threshold point (V_{t1}, I_{t1}) and holding point (V_h, I_h) are obtained readily. The discharging channel resistance can be derived from the I-V curve with the thermal-enhanced resistance effect observed clearly in this case. Choosing the silicon melting temperature as the critical failure temperature, the thermal failure threshold, or, the second breakdown point from which the ESD protection level is derived, can be obtained, which is however not clear in this case because of the large simulation steps used in simulation. Of course, one ought to be cautious in using these simulation data and calibration plays a critical role in ESD simulation. The essential portion of mixed-mode ESD simulation is to perform transient ESD simulation at circuit level, where an HBM model circuit is used to generate ESD pulses to stress the structure. For the targeted HBM protection levels of 2kV and 4kV, simulation shows that protection devices of 56μm wide and 111μm wide are needed, respectively. A group of simulated curves for a 4kV-passed case are given in Figure 8.15. Figure 8.15a shows a typical transient ESD I-V characteristic under a 4kV HBM pulse, where triggering threshold, holding point and discharging channel resistance can be obtained. Snapback behaviour is clearly observed. Readers are advised to pay attentions to the portion of I-V curve corresponding to the rising ESD pulse (solid line), but not the falling curve segment (dashed line). Triggering time,

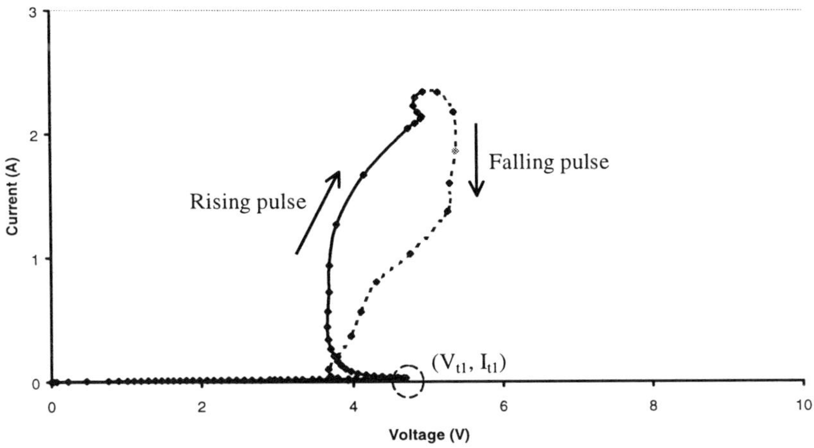

Figure 8.15a A typical I-V characteristic of ggNMOS from mixed-mode circuit transient simulation passed 4kV HBM ESD stress. Solid line corresponds to rising ESD pulse and dashed line relates to fall pulse [27].

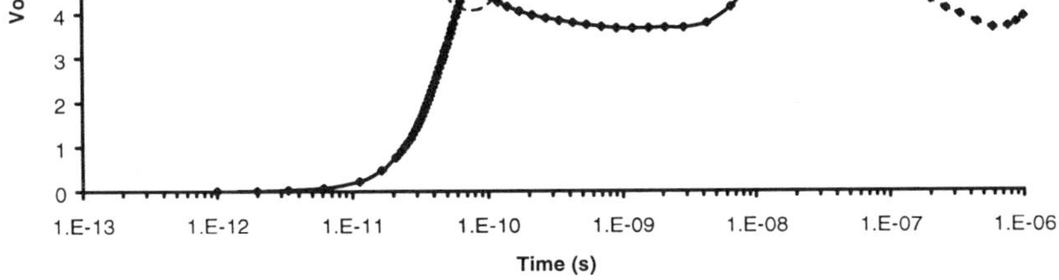

Figure 8.15b V-t curve for ggNMOS from circuit simulation provides response time information, which is critical to ESD protection operation [27].

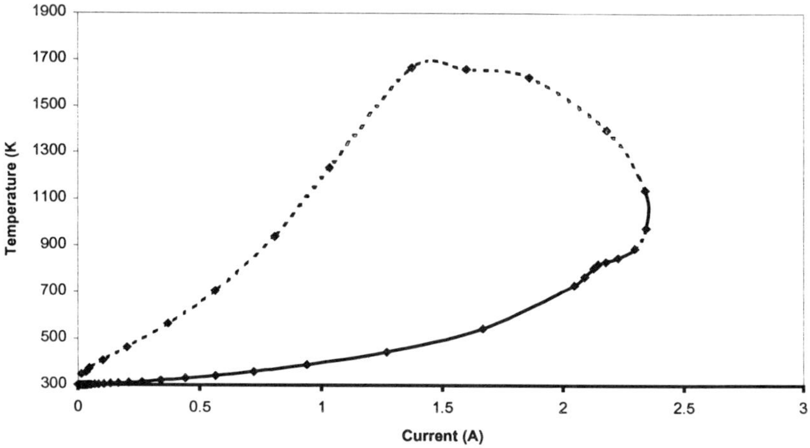

Figure 8.15c T_{max} -I curve shows temperature increase with ESD current in the rising pulse [27].

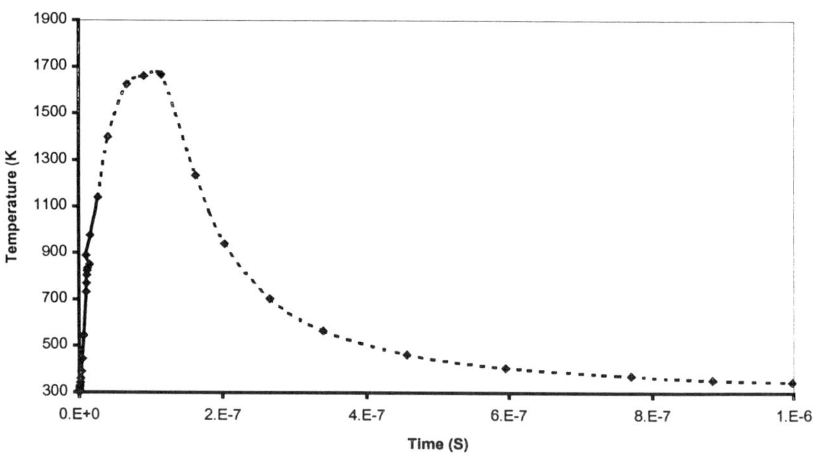

Figure 8.15d T_{max} -t curve of the ggNMOS under a HBM ESD stress shows monolithically increase in temperature during the rising ESD pulse. Temperature continuously increases as ESD pulse starts fall since heat dissipation takes time [27].

t_1, can be obtained from the time domain characteristic as shown in Figure 8.15b. Designing a short response time is critical to proper operation of an ESD protection structure because it must respond to a fast ESD transient quickly enough in order to provide ESD protection. Lattice temperature characteristics can be investigated in Figure 8.15c and d, where monotonically increased temperature is observed in the rising ESD pulse duration. As the ESD pulse starts to decrease, temperature continues to increase for a short period before decrease because of thermal dissipating procedure. ESD failure is demonstrated in Figure 8.16 for an under-sized ggNMOS structure stressed under 2kV. Figure 8.16a shows the I-V curve that is broken when lattice temperature exceeds the melting temperature and ESD simulation terminates and reports a failure message. The continuous increase of temperature is clearly observed in Figure 8.16b. Further details are depicted in the 2D and 3D maximum lattice temperature contours, T_{max}, in Figure 8.17 and Figure 8.18, where a hot spot at the drain junction is clearly observed that is a sub-surface heat source in silicon. The heating up in the metal interconnects at both source and drain sides are also observed in Figure 8.18. To illustrate the thermal breakdown threshold more clearly, Figure 8.19 shows an I-V characteristic from DC simulation for a ggNMOS ESD protection structure implemented in a 0.35μm CMOS process, which is a classic curve showing all key ESD operation points clearly [21]. Well, to avoid any misleading information, it is important for readers to understand that a hundred-percent prediction should not be expected for the time being and calibration is critical to using the ESD simulation methodology successfully.

8.5.2. Case 1: NMOS ESD Protection Structures in 0.8μm BiCMOS

With basic understanding of the procedures and results of mixed-mode ESD simulation, we proceed to demonstrate how to use the ESD simulation technique to direct practical ESD protection circuit design. The example used in this section is NMOS ESD protection structures as discussed in Chapters 3 and 4 implemented in an industrial 0.8μm BiCMOS technology [28]. The SCGS and DCGS values are set to be 2μm and 4μm, respectively, by simulation for maximum ESD performance. For a ggNMOS ESD protection structure, circuit level transient simulation, with the resulting curves shown in Figure 8.20, find a trigger voltage of $V_{t1} \sim 14.68V$ and a fairly short triggering time of $t_1 \sim 0.2ns$. The short response feature makes it a feasible solution to not only HBM but also the fast CDM and IEC ESD pulses. However, this trigger voltage is relative high for many applications.

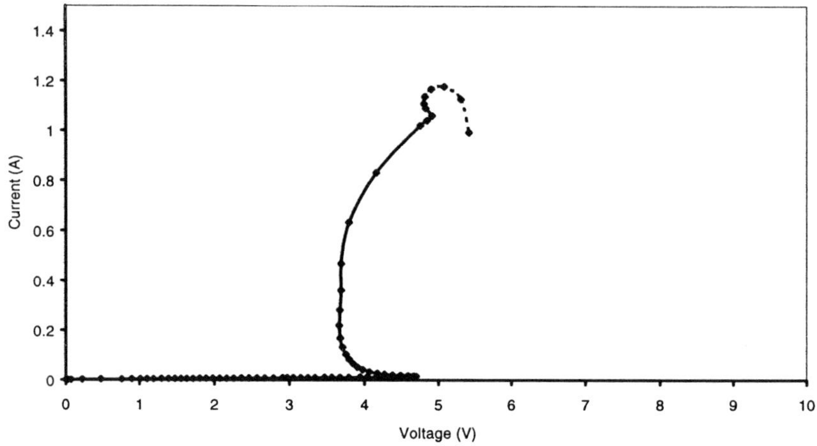

(a)

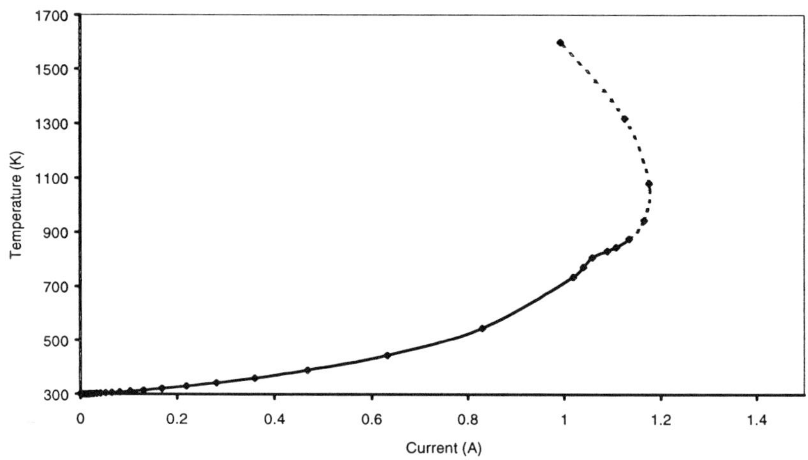

(b)

Figure 8.16 An under-sized ggNMOS failed a 2kB HBM ESD stress as simulation terminates upon temperature exceeding melting temperature: (a) I-V curve, (b) T_{max} -I curve [27].

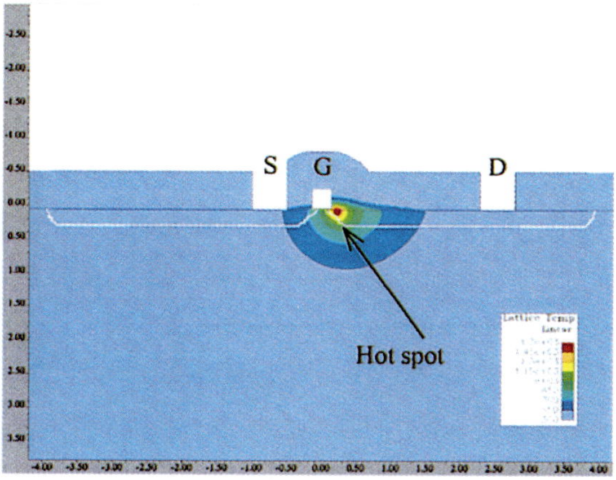

Figure 8.17 A 2D maximum lattice temperature contour shows a hot spot located at the drain junction under ESD stress [27].

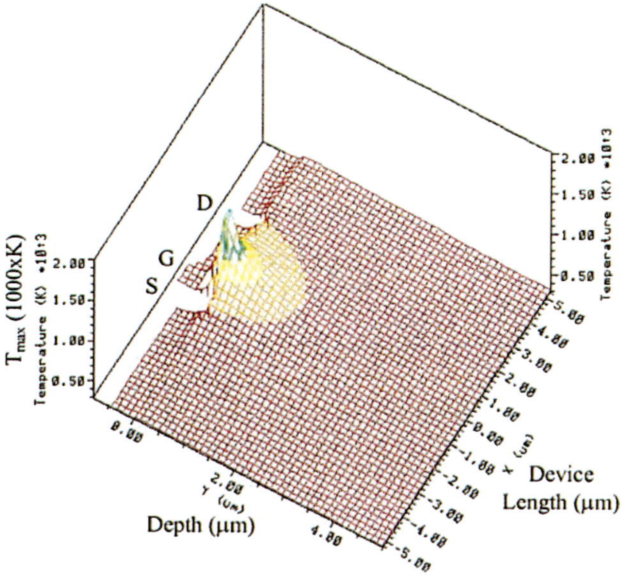

Figure 8.18 A 3D T_{max} contour indicates that the heat source is located in silicon surface. Temperature increase in the source and drain metal lines is also evident [27].

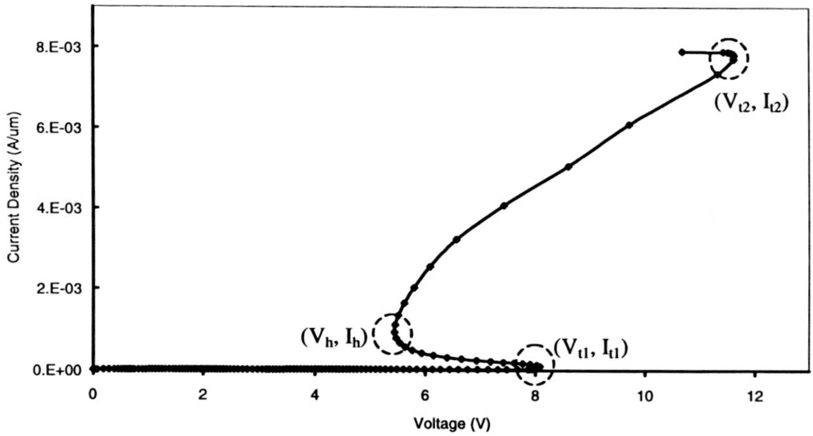

Figure 8.19 A classic I-V curve for a ggNMOS ESD protection structure, designed for a 0.35μm CMOS, provides clear information for triggering, holding and thermal failure threshold points [21].

Most critically, this trigger voltage is higher than that of the second breakdown voltage, V_{t2}, which will cause non-uniform turn-on of a multiple-finger ESD protection structure, as discussed in Chapter 4. As discussed before, one remedy to this high trigger voltage problem is to use a gcNMOS network to reduce V_{t1}. Transient ESD simulation helps to select the resistance and capacitance values as R=10kΩ and C=0.1pF for optimal performance. Typical ESD simulation results are shown in Figure 8.21 for the gcNMOS network. Figure 8.21a clearly shows that the trigger voltage is reduced substantially to V_{t1} ~ 7.54V without slowing down its response time as indicted in Figure 8.21b. The moderate pump-up of the gate bias, V_G, is readily observed in Figure 8.21d, which serves to accelerate the triggering of the protection structure as described in Chapter 4. These simulation results are confirmed by measurements, with typical parameters from both simulation and measurements summarized in Table 8.1. This practical ESD protection design example illustrates how mixed-mode ESD simulation can be used to predict ESD protection design performance, of course, not totally yet.

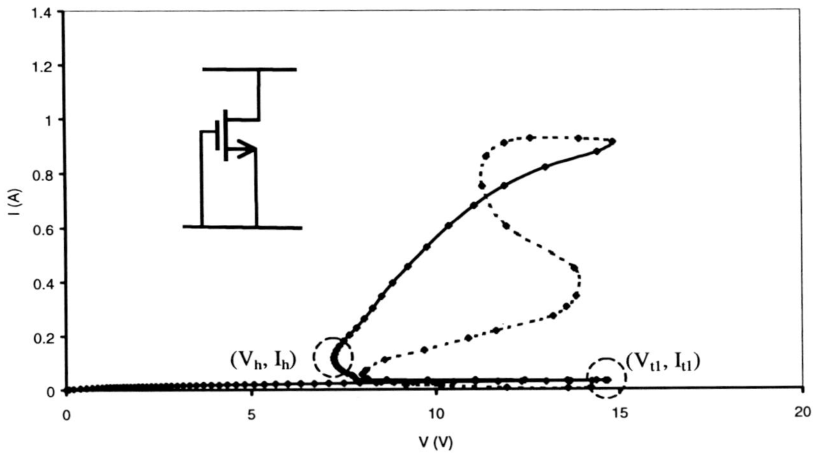

(a)

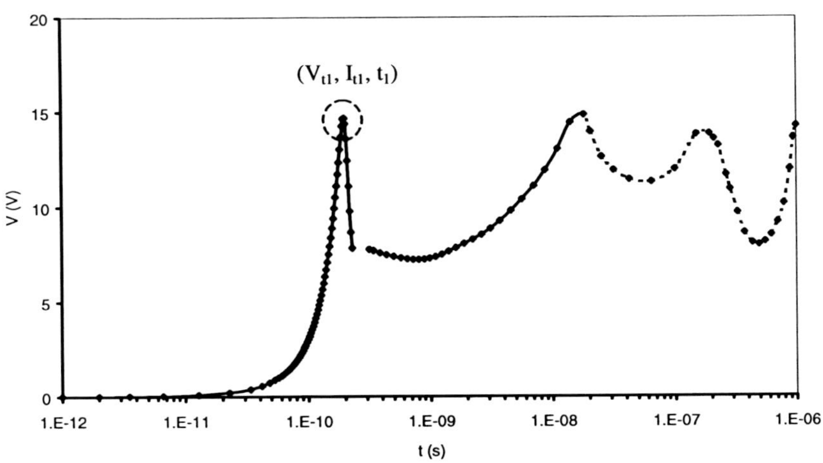

(b)

Figure 8.20 Transient simulation results for a ggNMOS device under 1.6kV HBM ESD pulse show (a) relatively high trigger voltage from I-V curve; (b) fast response from V-t curve [28] (*Reproduced here by kind permission of Elsevier Science and authors*).

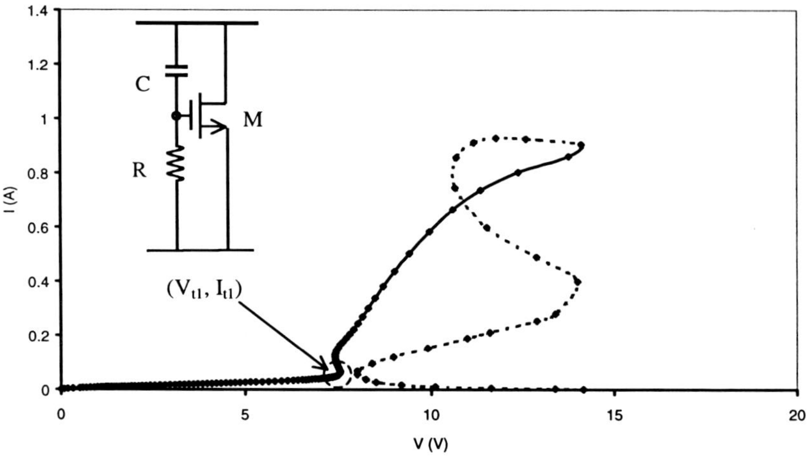

Figure 8.21a Simulated I-V curve for the gcNMOS protection network shows reduced V_{t1} [28] (*Reproduced here by kind permission of Elsevier Science and authors*).

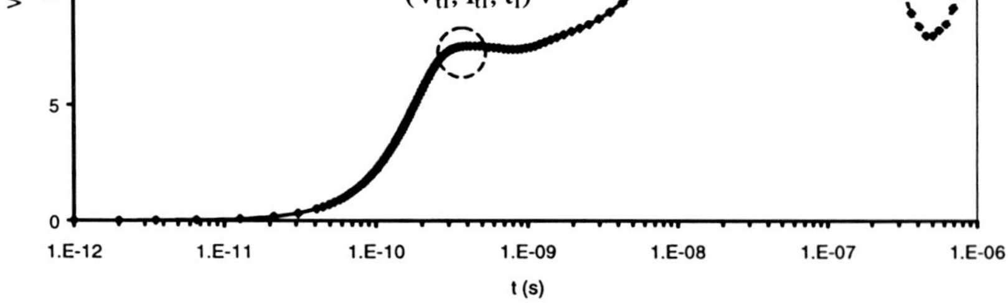

Figure 8.21b Simulated V-t curve shows no slowdown in response for the gcNMOS [28] (*Reproduced here by kind permission of Elsevier Science and authors*).

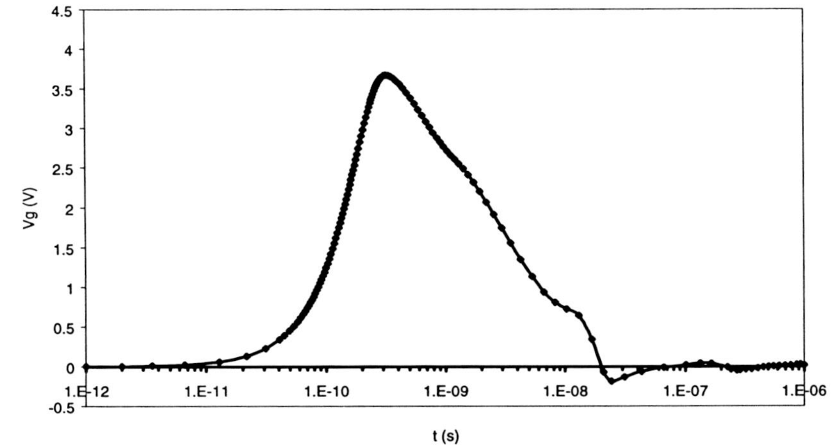

Figure 8.21c V_G is raised up for a short period for the gcNMOS network to reduce the V_{t1} [28] (*Reproduced here by kind permission of Elsevier Science and authors*).

Table 8.1 A summary of simulation and test data for the ggNOS and gcNMOS ESD protection structures shows reasonable matching between simulation and measurements. Reduction in V_{t1} is achieved via lifting up V_G. Good timing assures proper ESD protection operation [28] (*Reproduced here by kind permission of Elsevier Science and authors*).

Devices	ggNMOS		gcNMOS	
Parameters	Simulation	Test	Simulation	Test
V_{t1} (V)	14.68	12.56	7.54	6.66
t_1 (ns)	0.2	-	0.42	-
V_h (V)	6.92	6.48	7.41	6.08
V_G^{max} (V)	-	-	3.67	-
t_G^{max} (ns)	-	-	0.32	-

8.5.3. Case 2: MOS ESD Protection Circuit in 0.35μm CMOS

Mixed-mode ESD simulation can be used to verify ESD protection circuit design at circuit level. In a practical ESD protection circuit design example, a gcNMOS power clamp is implemented in a commercial 0.35μm CMOS technology for power line protection [21]. While the stand-alone gcNMOS ESD protection network functions by itself, one concern arises for using this ESD protection circuit in a special circuit block as shown in Figure 8.22, where two cascaded inverters are connected in parallel with the gcNMOS ESD protection unit. The design concern is with the integrity of gate oxide of the inverters when similar MOSFET transistors are used in both the inverters and the gcNMOS ESD protection structure. Assuming a logic low signal comes into the input of the first inverter, NMOS transistor M_{N1} remains off while PMOS transistor M_{P1} is turned on. Using an equivalent channel resistor to replace the on-PMOS, M_{P1}, the circuit is simplified to that shown in Figure 8.23. When an ESD pulse appears at the power bus, chance is that the gate oxide material of NMOS transistor in inverter two, M_{N2}, may be damaged if the gcNMOS ESD protection unit does not function efficiently. Careful design by simulation may eliminate this potential problem. Simulation results shown in Figure 8.24 and Figure 8.25 demonstrate this subtle design consideration. For the CMOS technology used in this work, the typical gate oxide breakdown voltage is BV_G = 8V with a worst-case value of 7V. Hence, a gate biasing of greater than 7V for a

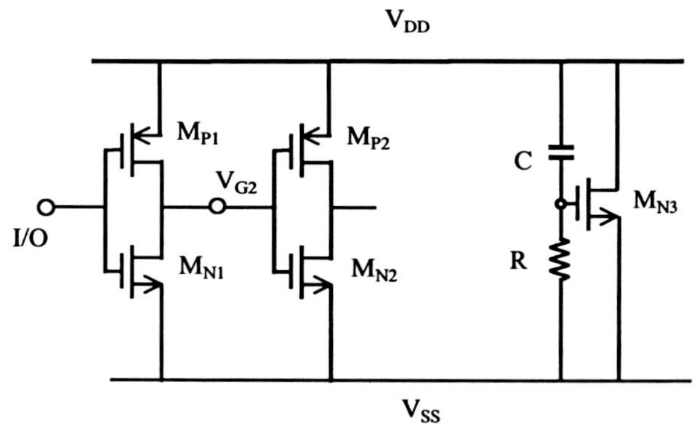

Figure 8.22 A special circuit block has two inverters connected in parallel with a gcNMOS power clamp [21].

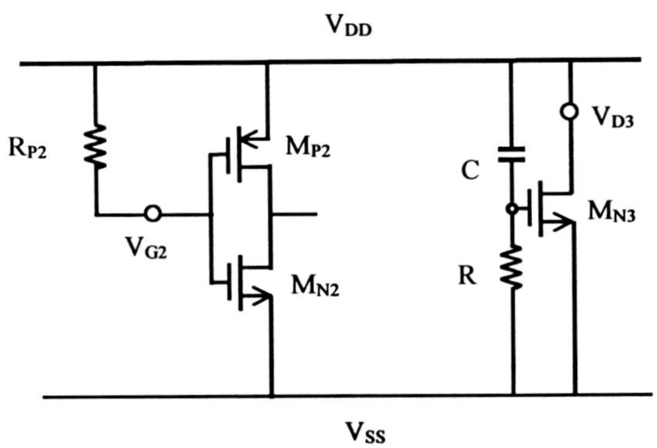

Figure 8.23 A simplified circuit for that in Figure 8.22 with a logic low input shows that the gate of transistor M_{N1} may be exposed to an ESD pulse [21].

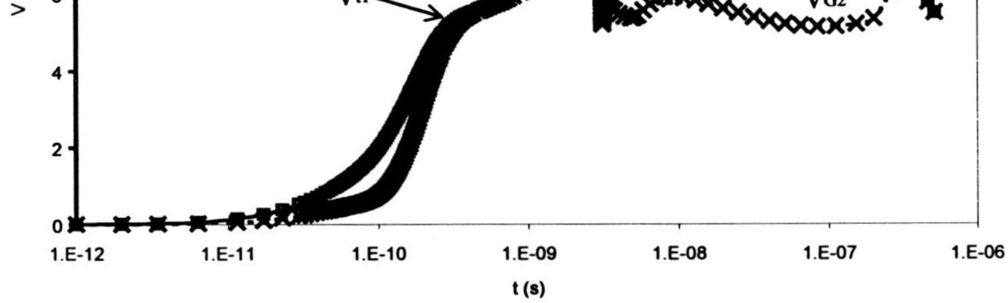

Figure 8.24 V_{D3} – t curve shows $V_{t1} < 6V$. V_{G2} – t curve shows a potentially risky region where V_{G2} briefly exceeds the worst-case gate breakdown voltage [21].

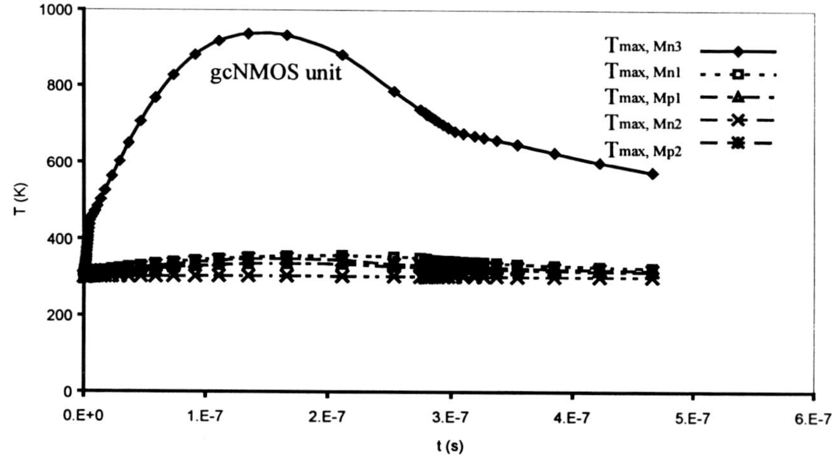

Figure 8.25 T_{max} – t curves show significant temperature increase inside the gcNMOS clamp only, indicating most ESD transient discharges through the gcNMOS structure [21].

substantial long duration may creates defects in gate oxide materials. In Figure 8.24, where terminal voltage of both the gcNMOS ESD clamp, V_{D3}, and gate bias of NMOS transistor M_{N2}, V_{G2}, are plotted in the time domain, a low trigger voltage of $V_{t1} < 6V$ is observed for the gcNMOS ESD protection unit. It is found that the gate voltage of the CMOS inverter never exceeds the 8V threshold. Detailed examination shows that although the gate bias V_{G2} briefly exceeds 7V by less than 1V, it lasts for shorter than merely 1ns. According to the process data, no defects in gate oxide are expected for such a short overstress. Therefore, the gcNMOS power clamp works properly in this case. As a further confirmation, the maximum lattice temperature curves given in Figure 8.25 show that significant increase in temperature is observed only in the gcNMOS ESD protection structure, but not in any other regular transistors in the inverters. This observation suggests that ESD transients mainly discharges into the gcNMOS protection structure.

8.5.4. Case 3: Metal Interconnect in ESD Protection Design

Metal interconnect design for ESD protection circuit is one of the most mysterious and confusing issues that has been overlooked most of the time. IC designers usually do not have any definitive and rational rules for metal interconnect design in ESD protection circuits. Commonly, a design rule book specifies a specific metal line width for metal interconnect design rule for ESD protection, typically, being 20μm. This 20μm-some metal line width design rule has been in place from 1μm processes to 0.18μm technologies for both traditional aluminium interconnects and new copper interconnect technologies. Another rule of thumb in design practice is to make metal interconnect in ESD protection section as wide as possible. While this passive metal design rule makes sense in old technologies where silicon damages are usually the main concern, one ought to challenge its validity in very deep sub-micron (VDSM) technologies. Apparently, a same, single, wide metal line width cannot be the optimal option for process technologies with different feature sizes. With the advances in designing silicon ESD protection structures, integrity of metal interconnect under ESD stresses becomes a design issue in VDSM technologies. A passive, blind-minded ESD metal interconnect design approach becomes questionable because it may either lead to ESD failure in metal interconnects due to inadequate line width, or, has negative parasitic influences on circuit performance due to excessive ESD metal coverage [27]. It is natural to expect a rational ESD metal design approach similar to the mixed-mode ESD simulation method used for silicon ESD protection structure design discussed previously. While a completely accurate simulation method for ESD metal interconnect design is still a research subject, one can benefit significantly by applying the mixed-mode ESD simulation approach described previously to ESD metal interconnect design. The idea is to treat metal materials, i.e., Al or Cu, as a semiconductor that can be handled by the existing TCAD tools by replacing relative material parameters by that for the metal. As an example, this simulation method was applied to practical ESD protection design in a commercial six-metal 0.18μm CMOS process technology having both Al and Cu interconnect modules to investigate, by simulation, the current handling capability of Al and Cu metal interconnects with different metal line widths and to evaluate the value of this ESD metal simulation approach [21, 27, 29]. Of course, the final goal is to design the metal interconnect for ESD protection rationally. The simulation results were confirmed by TLP test data in this work. Figure 8.26 shows the ESD robustness versus metal line width for Cu interconnects for six different metal layers from both simulation and measurements, which clearly shows the proportional relationship as expected. The significance of these curves is

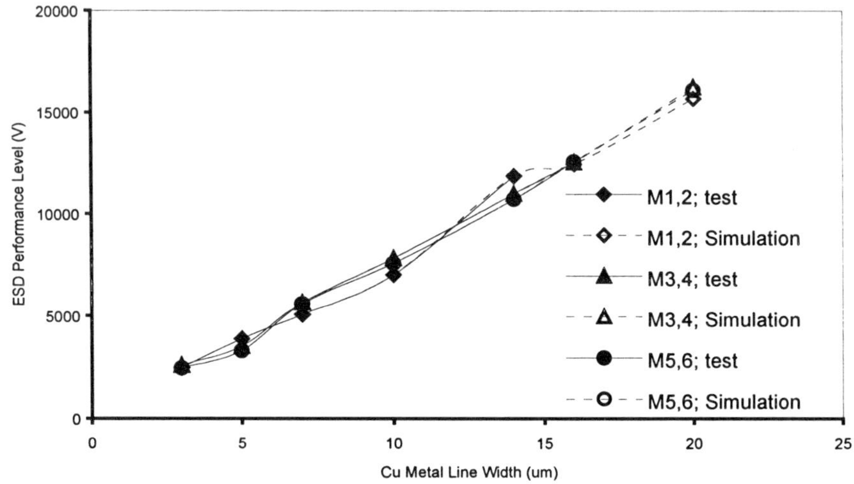

Figure 8.26 ESD robustness versus metal line width data for Cu interconnects show a proportional relationship and good matching between simulation and measurement [21, 27].

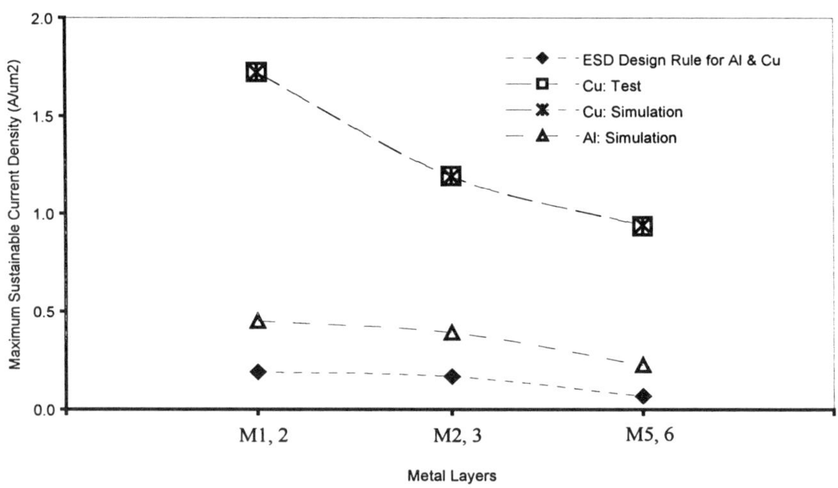

Figure 8.27 Current handling capability data derived from the process design rule book, simulation and measurement for six metal layers using either Al or Cu interconnects show the need for simulation in designing metal interconnects for ESD protection [21, 27].

the good matching between simulation and testing, indicating usefulness of this ESD metal simulation approach. Figure 8.27 gives a group of curves describing the maximum sustainable current densities for different metal layers derived from regular process design rule book, ESD simulation and testing for both Al and Cu. The data corresponding to the design-rule-book ESD metal design rules uses the 20μm metal line width rule specified in the CMOS process used, which is the same for both Al and Cu interconnect modules. These design-rule-book data are compared with the simulation and measurement results for Al and Cu interconnects, respectively. It is observed clearly that the all-fitting 20μm metal line width ESD design rule is very conservative compared with what really needed as suggested by both simulation and measurements. Good agreement between simulation and measurement is observed. Of course, it also confirms that Cu interconnect is superior to its Al counterpart in sustaining large current surges as expected. To waste a few more words in this matter, a designer should always take a scientific approach in practical design, instead of relying solely on experiences and/or other's successful stories. This example demonstrates how existing techniques can benefit designers if used wisely.

8.5.5. Case 4: A Dual-Direction ESD Protection Structure in BiCMOS

A useful ESD simulation-design methodology not only helps to design regular ESD protection structures specially tailored for specific process technologies and applications, but also plays a critical role in exploring new, novel ESD protection structures. Take the dual-direction ESD protection structure described in Chapter 5 as an example. The new structure, as shown in Figure 8.28 redrawn from Figure 5.8 for convenience, behaves like an SCR shunting devices in both directions under ESD stresses [30]. The new ESD protection structure was a completely new creature of mixed-mode ESD simulation and its functionality was verified thoroughly by simulation. The new structure was designed and implemented in a commercial 0.8μm BiCMOS process technology for mixed-signal IC chips. In the first step, DC simulation was conducted and the desired symmetric deep-snapback I-V characteristic was confirmed by simulation as shown in Figure 8.29. In the next step, its functionality was proved by transient ESD simulation using HBM ESD model circuit. Figure 8.30 shows its I-V characteristic in one direction from the anode to cathode terminal. With complete confirmation from mixed-mode ESD simulation, the design was implemented in silicon. Finally, measurements were conducted using both a curve tracer and a transient TLP tester, with its I-V characteristics illustrated in Figure 8.31 and Figure 8.32, respectively. Table 8.2 summaries typical simulation and testing

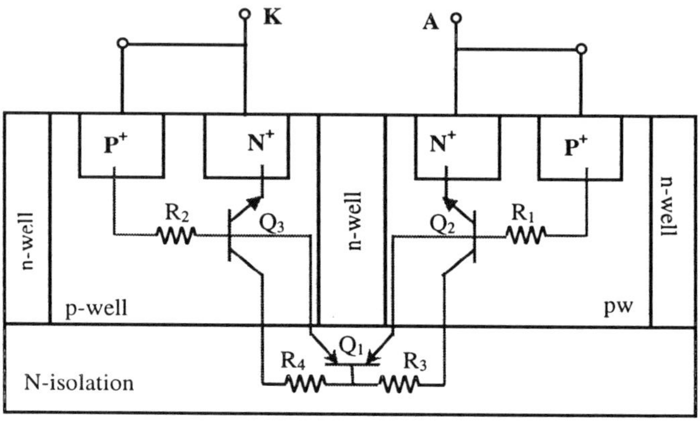

Figure 8.28 A dual-direction ESD protection structure in BiCMOS [30] (*Reproduced here by kind permission of IEEE and authors*).

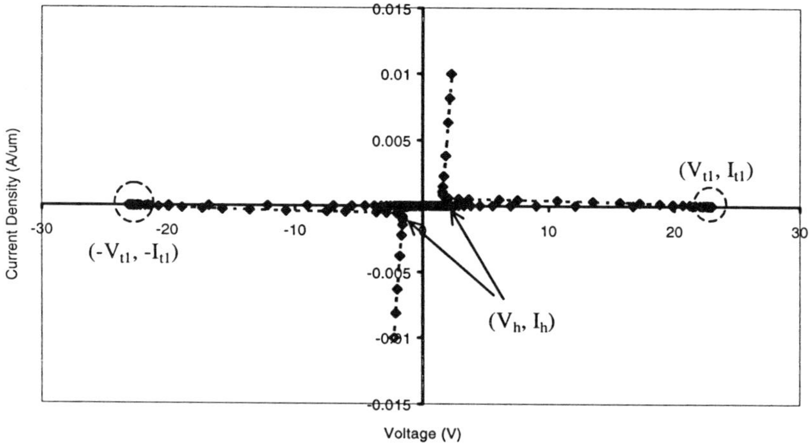

Figure 8.29 I-V characteristic from DC simulation for the dual-direction ESD protection structure shows the symmetric deep-snapback feature [30] (*Reproduced here by kind permission of IEEE and authors*).

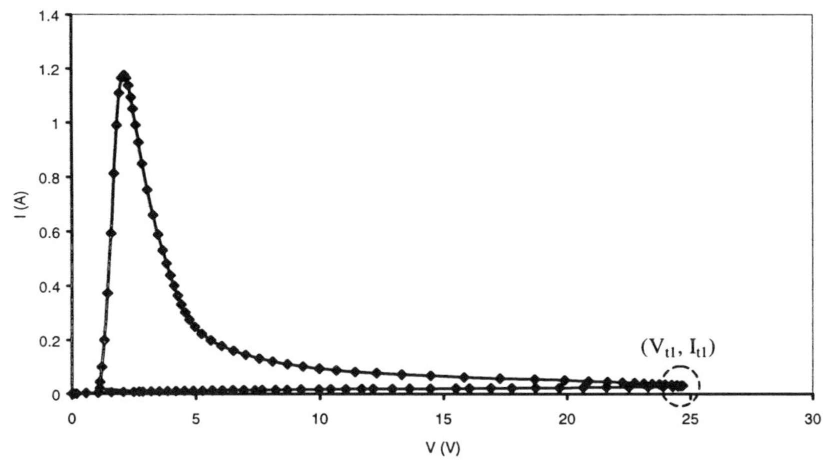

Figure 8.30 I-V characteristic from transient ESD simulation under a 2kV HBM ESD stress in one direction [30] (*Reproduced here by kind permission of IEEE and authors*).

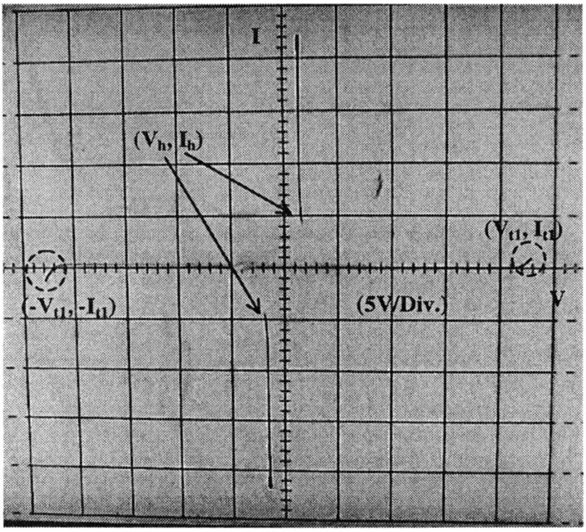

Figure 8.31 I-V curve from curve tracer testing agrees with that from DC simulation [30] (*Reproduced here by kind permission of IEEE and authors*).

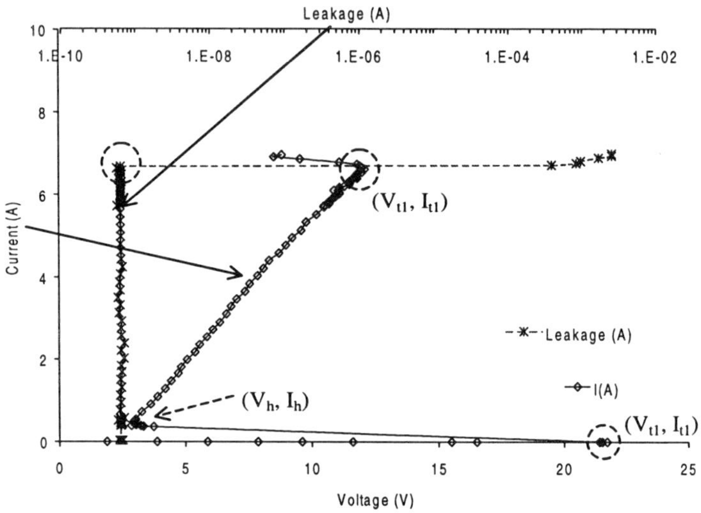

Figure 8.32 I-V curve from TLP test matches HBM simulation. Leakage current jumps up sharply at the thermal breakdown point [30] (*Reproduced here by kind permission of IEEE and authors*).

Table 8.2 Simulation and test data for the dual-direction ESD protection structure of 100μm wide include trigger voltage, holding voltage and discharging channel on-resistance [30] (*Reproduced here by kind permission of IEEE and authors*).

Parameters	Simulation	Measurement	
		Curve Tracer	TLP
V_{t1} (V)	23.32	22.55	21.75
V_h (V)	1.58	1.55	2.53
R_{ON} (Ω)	0.74	-	1.4

data for this new ESD protection structure. Reasonably good agreement is observed between the simulation and measurement results for this new design. Once again, to avoid any misleading, the above design examples are not given to showcase how successful those designs were and what a good matching between simulations and testing was achieved. Instead, they serve to argue that a rational simulation-design methodology should be used for predictive ESD protection circuit design and such approach is feasible for practical ESD protection design.

8.6. ESD PROTECTION DESIGN VERIFICATION

The main focus of this chapter is on ESD simulation methodologies, which serve to direct practical ESD protection circuit design in a predictive way. A good ESD simulation approach provides a systematic design procedure that works IC designers through the complex ESD protection circuit design processes. ESD simulation helps IC designers to prove the functionalities of ESD protection circuits at schematic level. However, ESD simulation by itself is not enough. A complete ESD design procedure should also include design verification to ensure design success. Theoretically, ESD design verification includes verifying the ESD protection circuit itself and confirming the full chip functionalities. Design verification for the ESD protection circuit portions includes traditional design rule checking (DRC) at layout level and layout versus schematic (LVS) checking. Simple physical spacing checking, series resistance checking and parasitic BJT device checking are the main focuses in such checking [31, 32]. However, such simple physical checking is not enough. A smart checking routine is highly desirable that should include extraction of irregular parasitic ESD-type devices, as well as extraction and comparison of critical terminal parameters. The reasons follow. Parasitic devices often compete against the designed ESD protection structures in shunting ESD transients, which may leads to pre-mature ESD failures. Most such parasitic devices have irregular patterns that make ESD protection circuit design extremely layout or geometry dependent. Further more, there are a large number of extracted ESD-type parasitic devices, of which most may not be able to play any role in ESD operation at all and should be excluded from any further verification work in order to conducting efficient design verification. The number reduction of parasitic ESD-type devices can be achieved by checking critical terminal parameters, such as, trigger voltage, holding voltage, discharging channel impedance, current gain, and so on. This smart checking procedure is referred to as *parametric checking*, which is advantageous to traditional DRC and LVS checking. On the other hand, full-chip design verification is necessary to ensure whole chip functionality. The reason is that, though a stand-alone ESD protection network may work well by itself, it may not function at chip level when being connected with the circuit being protected. In addition, any ESD protection structure may have negative influences on performance of the circuit being protected as well. Full-chip design verification serves to verifying functionalities of both the ESD protection networks and the circuit being protected at whole chip level. Currently, such sophisticated CAD software package is still under research [33].

8.7. SUMMARY

This chapter provides in-depth discussions on ESD simulation-design methodologies. As opposed to the traditional experience-based trial-and-error ESD protection design approaches, ESD simulation can be used in practice to guide ESD protection circuit design with the final goal of providing ESD protection design prediction. Several ESD simulation methods are available for ESD protection design including TCAD-based device level simulation and ECAD-based circuit level ESD simulation. The advantage of device level simulation is that it deals directly with semiconductor device physics equations that are essential to investigating ESD protection operation. However, it does not cover the whole picture at circuit level, which is critical in ESD protection operation. On the other hand, circuit level simulation takes care of the circuit level function nature of ESD protection operation and can be readily handled by ordinary IC circuit designers. Theoretically, as long as accurate device models for ESD-type devices are available, circuit level ESD simulation can be realized by using regular SPICE-type simulator. Unfortunately, such accurate ESD device modelling work is still under research currently. What makes accurate ESD device modelling difficult is the unique features associated with ESD behaviours, such as, very high current operation, avalanche breakdown, thermal-electro coupling effect, etc. In addition, since ESD protection performance is extremely layout geometry dependent with defects often occurring at the corners and edges of ESD protection structures, parallel section-wise device modelling techniques are essential to developing accurate ESD device models to ensure predictive ESD simulation at chip level. Recognising the multiple-level coupling effects in ESD protection operation, a mixed-mode ESD simulation-design methodology was developed for practical ESD protection circuit design. Since a mixed-mode ESD simulation approach includes the complex process-device-circuit-electro-thermal coupling effects, it enables IC designers to perform predictive ESD protection simulation in practical ESD protection circuit design practices. Several practical design examples are presented to demonstrate the value of the mixed-mode ESD simulation-design methodology. Finally, full-chip level ESD design verification is necessary for completely predictive ESD protection circuit design that includes smart physical design checking and evaluates the complex interactions between the ESD protection structures and the circuit being protected. Such whole-chip level ESD design verification software is currently under research. As the last remarks for IC designers, it is important to understand that previous design experiences are useful, however, they do not guarantee design success, or, most importantly, the first-time design success. ESD simulation must be used in practical ESD protection circuit design to increase success rate and to reduce development time. While one does not expect the same level of accuracy in ESD simulation as that enjoyed in regular circuit simulation currently, the benefit of using existing ESD simulation approaches in practical ESD protection circuit design is strikingly clear, well, provided proper calibration is exercised in ESD simulation.

REFERENCES

[1] Amerasekera, A. and Duvvury, C., *ESD in Silicon Integrated Circuits*, John Wiley & Sons, England, 1995.

[2] Beebe, S., "Characterization, Modelling, and Design of ESD Protection Circuits", *Stanford University Tech. Report,* No. ICL94-038, 1994.

[3] Luchies, J. M., "Electrostatic Discharge in Integrated Circuits; Testing and Protection", *Ph.D. Thesis*, Univ. of Twente, 1995.

[4] Voldman, S., Furkay, S. and Slinkman, J., "3D Transient Electrothermal Simulation of ESD Protection Circuits", in *Proc. EOS/ESD Symp.*, 1994, pp. 246-256.

[5] Rodriguez, J., Smayling, M., and Wilson, W., "ESD Circuit Synthesis and Analysis Using TCAD and SPICE", *IEDM Tech. Dig.*, 1998, pp97-100.

[6] Diaz, C. and Kang, S. M., *Modelling of Electrical Overstress in Integrated Circuits*, Kluwer Academic Publishers, 1994.

[7] Diaz, C., Kang, S. M. and Duvvury, C., "Circuit-Level Electricalthermal Simulation of Electrical Overstress Failures in Advanced MOS I/O Protection Devices", *IEEE Trans. on CAD*, vol. 13, no. 4, 1994, pp. 482-493.

[8] Sinha, S., Swaminathan, H., Kadamati, G. and Duvvury, C., "An Automated Tool for Detecting ESD Design Errors", *Proc. EOS/ESD Symp.*, 1998, pp208-217.

[9] Sze, S. M., *Physics of Semiconductor Devices, 2nd Ed.*, New York, Wiley, 1981.

[10] Dutton, R. and Yu, Z., *Technology CAD – Computer Simulation of IC Processes and Devices*, Kluwer Academic Publishers, Norwell, MA, 1993.

[11] Diaz, C., "*Modelling and Simulation of Electrical Overstress Failures in Input/Output Protection Devices in Integrated Circuits*", PhD Dissertation, University of Illinois at Urbana-Champaign, April 1993.

[12] Ramaswamy, S. and Kang, S. M., "iETSIM Version 2.0 Users Manual", *Tech. Rep., U. of Illinois at Urbana-Champaign*, 1996.

[13] Beltman, R., van der Vlist, H. and Mouthaan, A., "Simulation of Thermal Runaway during ESD Events", *Proc. EOS/ESD Symp.*, 1990, pp.157-161.

[14] Kurimoto, K., Yamashita, K., Miyanaga, I., Hori, A. and Odanaka, S., "An Electrothermal Circuit Simulation using an Equivalent Thermal Network for Electrostatic Discharge", *Proc. VLSI Symp. Tech.*, 1994, pp.127-128.

[15] Li, T., Tsai, C.-H., Rosenbaum, E. and Kang, S.-M., "Substrate Resistance Modelling and Circuit-Level Simulation of Parasitic Device Coupling Effects for CMOS I/O Circuits under ESD Stress", *Proc. EOS/ESD Symp.*, 1998, pp281-289.

[16] Lim, S. L., Zhang, X., Yu, Z., Beebe, S. and Dutton, R., "A Computationally Stable Quasi-Empirical Compact Model for Simulation of MOS Breakdown in ESD-Protection Circuit Design", *Proc. Int'l Conf. Semicon. Processes and Devices*, 1997, pp.161-164.

[17] Tu, R., King, J. C., Shin, H. and Hu, C., "Simulating Process-Induced Gate Oxide Damage in Circuits", *IEEE Trans. Electron. Dev., Vol. 44, No. 9,* September 1997, pp.1393-1400.

[18] Satake, H. and Toriumi, A., "Dielectric Breakdown Mechanism of Thin SiO2 Studied by the Post-Breakdown Resistance Statistics", *IEEE Trans. Electron. Dev., Vol. 47, No. 4,* April 2000, pp741-745.

[19] Wang, A. Z., Feng, H. G., Gong, K., Zhan, R. Y. and Stine, J., "On-Chip ESD Protection Design for Integrated Circuits: an Overview for IC Designers", *J. Microelectronics, Elsevier Science, Vol. 32/9,* August 2001, pp.733-747.

[20] Wang, A. and Tsay, C., "A Novel Design Methodology Using Simulation for on-chip ESD Protection for Integrated Circuits", *Proc. IEEE Intl. Conf. Solid-State & IC Technology*, 1998, pp.509-512.

[21] Feng, H. G., "A Mixed-Mode Simulation-Design Methodology for on-Chip ESD Protection Design", *MS Thesis*, May, 2001.

[22] *TSUPREM-4, 1D/2D Semiconductor Process Simulation Software*, Avant! Corp.

[23] *MEDICI, 2D Semiconductor Device Simulator*, Avant! Corp.

[24] *TAURUS, 1D/2D/3D Semiconductor Process and Device Simulation Software*, Avant!. Corp.

[25] *DESSIS 6.0, 1D/2D/3D Device and System Simulator*, Integrated Systems Engineering, AG.

[26] Stricker, A., Mettler, S., Wolf, H., Mergens, M., Wilkening, W., Gieser, H. and Fichtner, W., "Characterization and Optimisation of a Bipolar ESD-Device by Measurements and Simulations", *Proc. EOS/ESD Symp.,* 1998, pp. 290-300.

[27] Feng, H. G., Gong, K. and Wang, A., "ESD Protection Design Using Copper Interconnects: More Robustness and Less Parasitics", *SRC Publication: 2000 Publications in Copper Design Challenge, Pub. P000375,* 17-Mar-2000.

[28] Wang, A., Tsay, C. and Deane, P., "A Study of NMOS Behaviours under ESD Stress: Simulation and Characterization", *Microelectronics Reliability, 38, Elsevier* Science, 1998, pp.1183-1186.

[29] Feng, H. G., Gong. K. and Wang, A., "A New Integrated Metal-Semiconductor Simulation Methodology for on-Chip Electrostatic Discharge Protection Design Optimisation ", *Proc. IEEE/IFIP World Computer Congress, IC Design Automation Conference*, 2000, p81 – 85.

[30] Wang, A. Z. and Tsay, C., "On a Dual-Polarity Electrostatic Discharge Protection Structure", *IEEE Trans. Elec. Dev., Vol. 48, No. 5,* May 2001, pp.978-984.

[31] Sinha, S., Swaminathan, H., Kadamati, G. and Duvvury, C., "An Automated Tool for Detecting ESD Design Errors", *Proc. EOS/ESD Symp.*, 1998, pp208-217.

[32] Li, T., Huh, Y. and Kang, S., "Automated Extraction of Parasitic BJTs for CMOS I/O Circuit under ESD Stress", *Int'l Reliability Workshop Final Report,* 1997, pp.103-109.

[33] Zhan, R. Y., "ESDcat: a CAD Package for Full-Chip ESD Design Verification", PhD thesis research, on going, October 2001.

Chapter 9

ESD – CIRCUIT INTERACTIONS

A successful on-chip ESD protection circuit design should take a system approach in order to address the mutual influences between the ESD protection networks and the core circuit being protected. This chip level ESD-circuit interaction problem is discussed in this chapter. Design examples are presented to demonstrate the importance of this problem.

9.1. CHIP-LEVEL ESD PROTECTION DESIGN

So far, ESD protection circuit design has been treated as a clean task of developing functional on-chip ESD protection structure itself. It has been a fairly common practice that an IC circuit designer issues the request for an ESD protection structure for a new IC product and an ESD protection technologist designs the ESD protection unit accordingly. The designed and verified ESD protection network is then delivered back to the IC designer to construct the whole IC chip. This approach sounds politically correct in general. However, very often, it ends up with a failed chip, even though the ESD protection structure itself has been carefully designed and thoroughly tested. Sometimes, the very same ESD protection structure may have already been used successfully in a different IC product in the same process technology before. A common debugging procedure usually calls the ESD technologist to investigate what is wrong with the ESD protection unit. In fact, the true culprit here is often the mutual influences between the ESD protection network and the core circuit being protected at chip level. The reality is that the whole chip, consisting of both the core circuit and the ESD protection units, should be treated as a single integrated circuit, where the two entities interact with each other in operation. On one hand, a functional stand-alone ESD protection structure may not work when connected to the core circuit because the core circuit may affect the ESD protection unit substantially. This is referred to as the circuit-to-ESD influence. On the other hand, ESD protection is not delivered for free. It is easy to image that any extra devices, i.e., the ESD protection structures, will affect the overall performance of the core circuit. This is defined as the ESD-to-circuit influence. Hence, this ESD-circuit interaction must be taken into consideration in any real chip design [1-3]. An ESD protection circuit design task ought to be considered in a system design manner in order to guarantee full chip functionality. This ESD-circuit interaction problem is a very complicated design issue. However, with proper cares exercised, it is a very doable work.

The first thing of concern to IC designers is associated with the process design rules (DR). It is common that a DR book of a specific process technology defines fixed layout design rules for every layer. While these fixed design rules are essential in circuit design and do not normally cause any problems in IC design, it often poses great constraints on ESD protection circuit design. For example, ESD simulation may suggest a narrower-than-the minimum channel length for a ggNMOS ESD protection structure in order to obtain larger current gain for the parasitic lateral bipolar transistor under an ESD stress. Similarly, if the ggNMOS ESD protection structure is connected in parallel with a minimum size NMOS transistor on a chip, to ensure the ggNMOS be turned on well before the minimum NMOS device does and takes all the ESD transient, the channel length of the ggNMOS must be designed narrower than that of the minimum NMOS device. However, normal design rules certainly disallow a designer doing so, even though the minimum gate width design rule may not affect ESD protection devices at all. Nevertheless, this design rule impact on ESD protection circuit design can often be relaxed with assistance from device engineers. On the other hand, the complex ESD-circuit interaction problem is not that straightforward and demands much more efforts from IC designers. The two aspects of ESD-circuit interaction are discussed in the following sections.

9.2. CIRCUIT-TO-ESD INFLUENCES: PRE-MATURE ESD FAILURES

Often, chip level ESD zapping tests lead to earlier ESD failures even though individual ESD protection structures demonstrate much higher ESD protection levels. The reason is that, when an ESD protection structure is connected to the core circuit being protected, ESD protection performance may be affected by the internal circuit being protected at chip level. This negative circuit-to-ESD influence is a very common root cause to pre-mature ESD failure of IC chips. Two typical types of circuit-to-ESD influence problems are discussed here.

The first type of circuit-to-ESD influence is associated with parasitic shunting links inside the core circuit. What happens here is that some internal devices or parasitic structures inside the core circuit may be turned on before the designed ESD protection unit is triggered during an ESD event and forms a parasitic ESD discharging channel. Even if an internal parasitic device is triggered along with the dedicated ESD protection structure, if the shunting resistance of the parasitic internal device is lower than that of the designed ESD protection structure, substantial ESD transient current will flow through the internal parasitic channel. Unfortunately, since none of the internal devices are designed to handle large ESD currents, ESD damage will occur inside the core circuit even though individual ESD protection structures are robust enough by themselves. Hence, lower turn-on voltage of any internal devices and/or the discharging competition between the intentional ESD protection network and the parasitic internal shunting device may eventually cause pre-mature ESD failures at chip level. For example, in the ESD protection scheme shown in Figure 9.1, a ggNMOS ESD protection structure is used for I/O pad protection. At chip level, it is very possible that the ggNMOS ESD protection structure, M_{ESD}, may be electrically connected

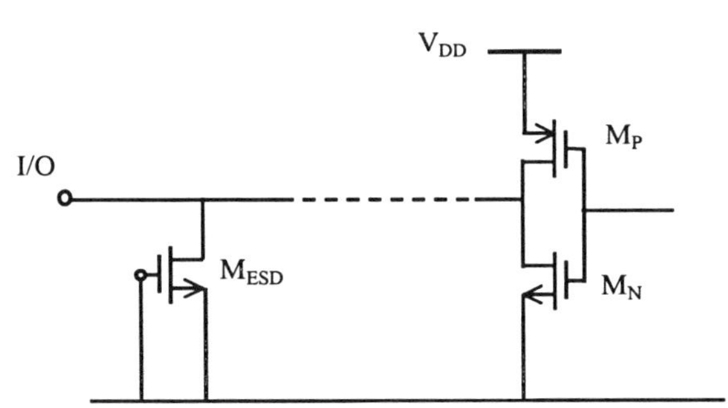

Figure 9.1 In a case where a ggNMOS ESD protection structure is used for I/O protection, if the M_{ESD} is connected to an internal NMOS transistor, M_N, in a dotted way, the ESD transient may discharge through the M_N instead of the M_{ESD} if the former device has lower trigger voltage and/or lower discharging resistance.

to the drain of an internal NMOS transistor, M_N, through a direct/indirect link. Since the two NMOS transistors, M_{ESD} and M_N, are effectively in parallel, the internal transistor, M_N, will be turned on along with the ggNMOS ESD protection device, M_{ESD}. In a worst case, if the ggNMOS transistor is not a minimum size device, which is common in ESD protection design, the internal transistor, M_N, may actually be turned on prior to the M_{ESD}. Consequently, the internal device, M_N, will take a fair amount of ESD transient or even the total ESD current. Early ESD damage in M_N is therefore expected in this case. Several measures can be taken to address this discharging competition problem in design. The ggNMOS ESD protection device, M_{ESD}, should be designed to have a lower trigger voltage than that of the internal NMOS, M_N. This can be done readily by using different diffusion layers in the MOSFET transistors. If the two MOSFET transistors are of the same type, the channel length of M_{ESD} should be designed shorter than that of the internal NMOS M_N so that the M_{ESD} has a lower discharging resistance compared to the M_N. This way, even though the internal NMOS, M_N, may be turned on during an ESD event, most ESD current will still flow into the ggNMOS ESD protection device, M_{ESD}. Therefore, the internal ESD defect in M_N will not be expected.

In a second design example as shown in Figure 9.2, a NMOSFET switch type ESD protection power clamp unit, as depicted in Figure 4.24, is used to

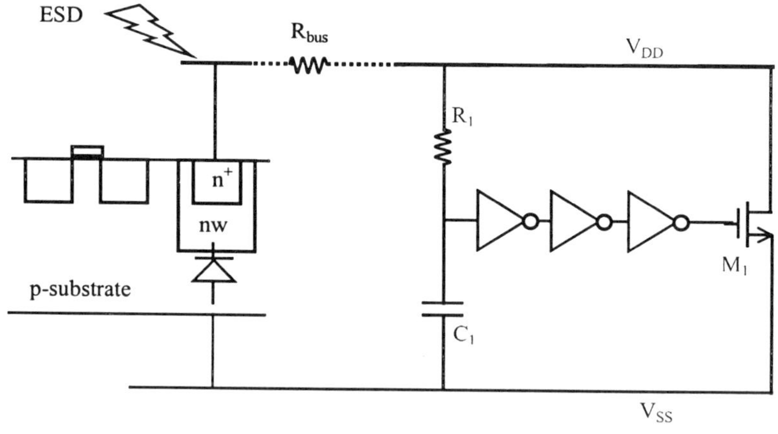

Figure 9.2 Improper placement of ESD power clamps on a chip may cause pre-mature ESD failure due to internal parasitic discharging path.

protect ESD surges in a power bus. As discussed before, placement of these power clamps on a chip is very important in layout design. Suppose the power clamp is not placed properly and an ESD transient is injected into the power bus at a V_{DD} pad located far away from the power clamp with the long distance being reflected by a noticeable bus resistance, R_{bus}. If it happens that there is an n-well guard-ring located near the V_{DD} pad, then a reverse-biased n-well/p-substrate diode is connected in parallel with the power clamp, remotely through the R_{bus}, because the n-well guard-ring and the p-substrate should be connected to V_{DD} and V_{SS}, respectively. In this scenario, since the ESD transient comes into the V_{DD} pad far away from the power clamp and there is a substantial bus resistance existing, the triggering of the power clamp will be significantly slowed down. Hence, there is a big chance that the parasitic n-well/p-substrate diode may be reverse broken down prior to the power clamp, therefore leading to ESD damage in the n-well guard-ring region. It is therefore very important to place the power clamps properly on a chip to avoid similar pre-mature ESD failures.

The next example illustrates how to use the mixed-mode ESD simulation to investigate the troublesome ESD discharging competition problem discussed previously in design phase. In a circuit block shown in Figure 9.3, a gcNMOS power clamp is used to protect the power bus. A CMOS inverter unit is located nearby and connected with the gcNMOS power clamp unit in

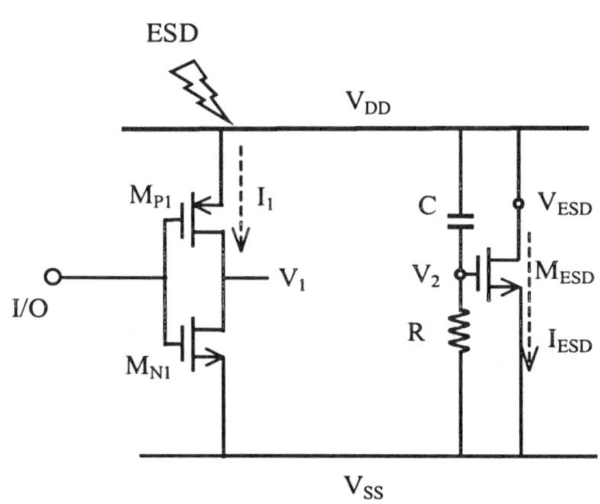

Figure 9.3 In an ESD protection scheme, where a gcNMOS power clamp is connected in parallel with an internal inverter, discharge competition may be a design concern at a logic "H" input [1].

parallel. This ESD protection circuit is implemented in a commercial 0.35μm CMOS process technology, where the gcNMOS features a W/L ratio of 80μm/0.35μm for the M_{ESD} and the internal inverter has the minimum sizes, i.e., W/L=1.5μm/0.35μm for the M_N and W/L=2μm/0.35μm for the M_P [1]. In operation, assuming a logic "H" signal comes to the input of the inverter, it will turns on the NMOS, M_N, and turns off the PMOS, M_P, in normal situation. However, if a positive ESD transient is injected into the power bus at this moment, the PMOS, M_P, is actually forced on by the large positive ESD pulse. Therefore, a discharge competition situation is created where both M_N and M_P of the inverter are turned on, along with the gcNMOS structure during the ESD event. Hence, it is likely that pre-mature ESD failure may occur in the inverter unit, even though the gcNMOS power clamp is triggered off properly. This discharge competition situation can be studied by performing mixed-mode ESD simulation following the approach described in Chapter 8. The simulation results are given in Figure 9.4 and Figure 9.5. Figure 9.4 shows the V-t and I-t characteristics obtained at several critical nodes and circuit branches. It is found that the NMOS ESD

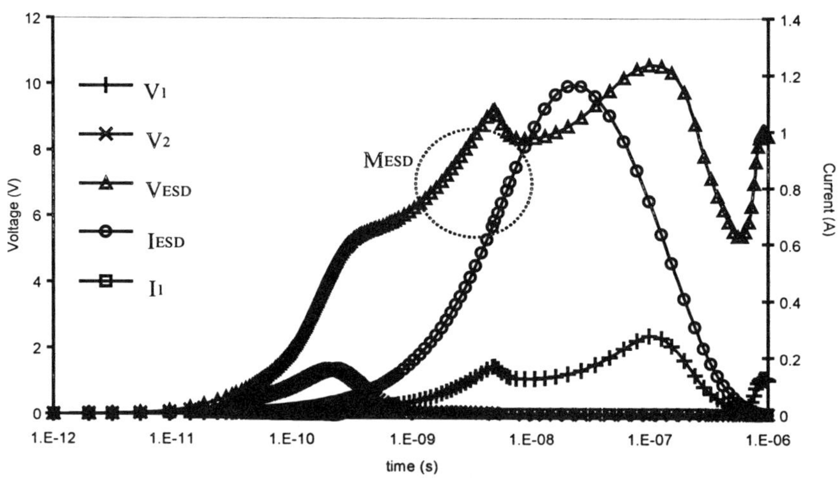

Figure 9.4 Mixed-mode ESD simulation shows that the M_{ESD} is triggered and discharges almost all the ESD current with negligible leakage current shunting into the inverter transistors [1].

protection structure, M_{ESD}, is triggered off during the ESD transient as expected as shown by the V_{ESD} – t curve. In the mean time, even though the inverter transistors are turned on by the ESD transient, almost all ESD current still flows into the gcNMOS structure (I_{ESD} – t curve) with negligible current leaking into the inverter transistors (I_1 – t curve). As a further confirmation, Figure 9.5 shows clearly that significant heat generation is only observed in the ESD protection transistor, M_{ESD}, but not in the inverter transistors, as depicted by the maximum lattice temperature curves. Therefore, the mixed-mode ESD simulation results indicate that the potential discharge competition risk is actually not a real problem in this design.

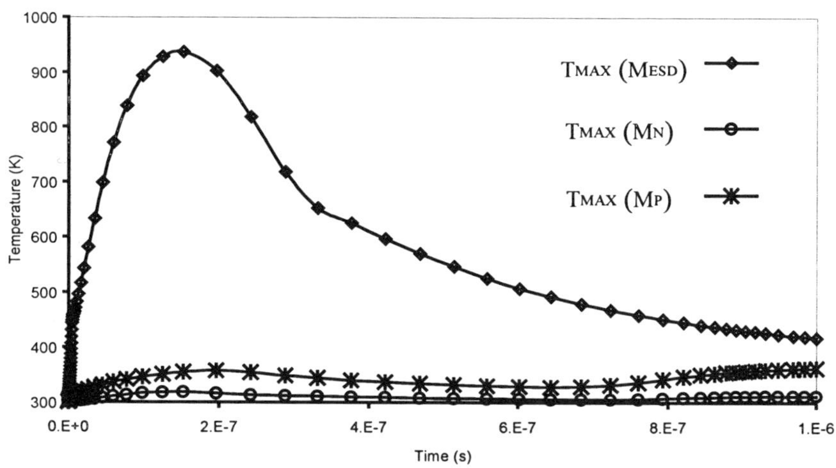

Figure 9.5 Mixed-mode ESD simulation shows that the M_{ESD} protection structure discharges the ESD transient, resulting in significant increases in its maximum lattice temperature, T_{MAX} (M_{ESD}) [1].

The second type of circuit-to-ESD influence is high-frequency signal induced accidental triggering of ESD protection units at chip level. This early turn-on problem is particularly a severe problem in mixed-signal and RF applications. As discussed previously in Chapter 5, a large dV/dt value of a high-frequency RF signal or a large magnitude digital noise in mixed-signal circuit results in significant displacement current, which may lead to accidental triggering of an ESD protection structure under normal operation conditions. This displacement current induced early turn on effect makes ESD protection circuit design a new challenge in high-frequency applications.

All the above design examples confirm that significant circuit-to-ESD influence could be a real design problem, which should be carefully addressed in practical ESD protection circuit design.

9.3. ESD-TO-CIRCUIT INFLUENCES: CIRCUIT PERFORMANCE DEGRADATION

Recognizing that ESD protection structures are extra entities to the core circuits, it is easy to expect some negative impacts on the core circuit performance from the ESD protection networks and design trade-offs are to be struck between the ESD protection sub-circuit and the core circuit being protected in practical design. While the circuit-to-ESD influence concept is commonly accepted in design community, the ESD-to-circuit influence does not get much attention yet, sort of because IC designers normally struggle to meet the ESD protection requirements and ESD protection structures are usually designed by non-circuit-designer ESD specialists. However, nowadays, the ESD-to-circuit influence becomes a real concern in designing high-frequency and high-density VDSM ICs, particularly for advanced RF and mixed-signal (MS) chips. The main ESD protection design trade-off in advanced RF and MS applications is to achieve high ESD protection level (e.g., beyond 4kV) and to maintain very low parasitic effects that exist in all ESD protection units. The ESD-to-circuit influence is certainly a multiple-fold problem, of which the key issues are associated with the substantial silicon consumption of the ESD protection structures and the ESD-induced parasitic effects. The silicon consumption of ESD protection structures is an obvious problem that becomes more severe in advanced IC chip design because the demand for ESD protection level is getting higher, which is normally achieved by using larger ESD protection structures. Another disadvantage of using large size ESD protection structures, such as a traditional NMOS ESD protection structure, is that it makes full chip layout design extremely difficult. The solution to this large-size problem is to use novel, compact ESD protection structures that can achieve the same or better ESD robustness with the same or smaller device size. In addition, innovative layout design of an ESD protection structure, such as a bonding pad oriented protection structure as discussed in Chapter 7, is very advantageous to full chip layout design. The inevitable ESD-induced parasitic effects include parasitic RC delay and noises. The problem caused by the ESD-induced parasitic RC delay, due to parasitic capacitance, C_{ESD}, and resistance associated with ESD protection structures, is obvious because it affects clock speed, signal integrity and other general circuit specifications. The ESD-induced noise effect includes both substrate noise coupling due to the C_{ESD}

and ESD self-generated noises. These ESD-induced parasitic effects may be killing factors for advanced RF and MS ICs in VDSM regime. In this section, a few practical design examples are presented to elaborate the ESD-to-circuit influence problem.

The design example used in this Section is a case study that uses three different ESD protection structures and three different circuits to demonstrate the significance of the ESD-to-circuit influence effect [2, 4, 5]. The study was conducted using a commercial 1.5V, six metal, 0.18μm CMOS process technology with both Al and Cu interconnect modules.

To study the degree of ESD-to-circuit influence due to different ESD protection structures, three types of ESD protection structures, i.e., a large-size traditional ggNMOS structure (as shown in Figure 3.17), a dual-direction SCR-type structure (as shown in Figure 5.8) and an all-in-one protection SCR-type structure (as shown in Figure 5.11) as described in Chapters 3 and 5 were used in the case study. These three ESD protection structure were named as ESD1, ESD2 and ESD3 in this study. To account for ESD metal interconnect design factors as well, the design were implemented in both Al and Cu interconnect modules using the rational ESD metal line design approach by simulation as described in Chapter 8, which found that around 30% reduction in ESD metal line coverage can be achieved when using Cu as compared to using Al interconnect for the same ESD protection level. This reduction in ESD metal coverage in using Cu interconnect translates into less parasitic capacitance associated with ESD metal lines when using Cu as opposed to using Al interconnect technology. To ensure meaningful study, the same ESD protection specification of 4kV HBM ESD robustness was used in the study. To achieve the same 4kV ESD protection level, different sizes for these ESD protection structures were used according to their individual performance level. The design goal was set to provide complete ESD protection for an I/O pad using the complete ESD protection scheme for each pad as described in Chapter 5 for these three ESD protection structures, ESD1, ESD2 and ESD3, corresponding to the diagrams of Figure 5.5, Figure 5.10 and Figure 5.12. For readers' convenience, these three ESD protection schemes are redrawn here as Figure 9.6, Figure 9.7 and Figure 9.8, respectively. To estimate the overall ESD-induced parasitic capacitances for these three different ESD protection structures, the parasitic capacitance models for ESD1, ESD2 and ESD3 described before are used to calculate the parasitic capacitances, C_{Si}, associated with the bare silicon ESD protection structures, taking into consideration of the numbers of ESD protection devices needed for complete ESD protection as illustrated in Figure 9.6, Figure 9.7 and Figure 9.8. The

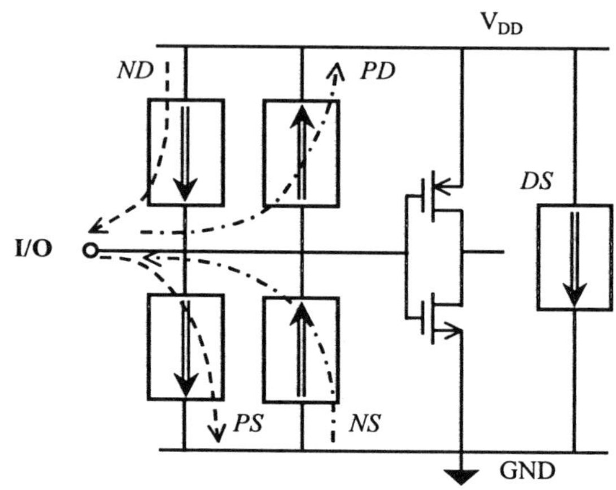

Figure 9.6 A complete ESD protection scheme using traditional one-direction ggNMOS ESD protection device, ESD1, needs multiple ESD protection units for each pad.

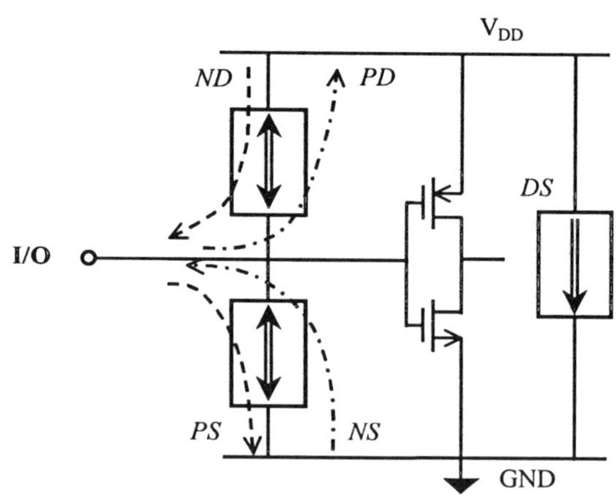

Figure 9.7 A complete ESD protection scheme using a dual-direction ESD protection structure, ESD2, need two units per pad.

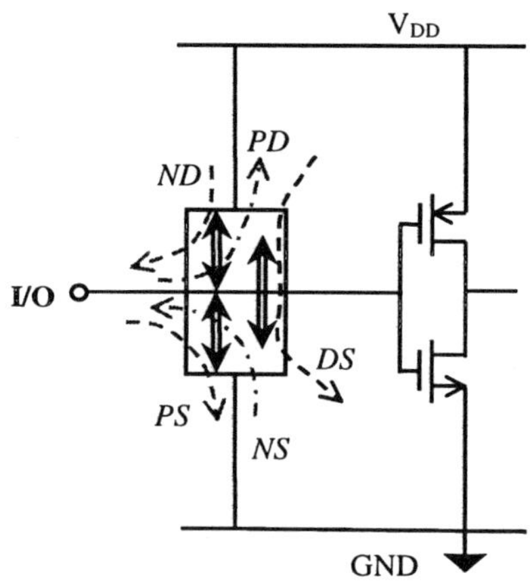

Figure 9.8 A complete ESD protection scheme using the all-in-one ESD protection structure, ESD3, needs only one such device and is very area-efficient.

parasitic capacitances associated with the ESD metal lines, C_M, include the metal-metal, metal-poly and metal-substrate capacitances. The overall parasitic ESD capacitances, C_{ESD}, are the sum of C_{Si} and C_M. The calculated ESD-induced parasitic capacitance data are summarized in Table 9.1 for C_{Si} and C_M, and Table 9.2 for the overall C_{ESD}. Several observations follow. First, sizeable parasitic C_{Si} exists in traditional ggNMOS ESD protection structure, ESD1, which was reduced substantially when using the two new ESD protection structures, i.e., ~83% reduction in ESD2 and a further ~22% decrease in ESD3 from ESD2. Secondly, considerable parasitic C_M was produced by ESD metal coverage in ESD1 in Al technology. However, use

Table 9.1 Estimated C_{Si} and C_M for ESD1, ESD2 and ESD3 in Al and Cu interconnects for 4kV full ESD protection [4].

Structures	ESD1		ESD2		ESD3	
C_{Si} (pF)	0.54		0.09		0.07	
C_M (pF)	Cu	Al	Cu	Al	Cu	Al
	0.30	0.43	0.029	0.041	0.019	0.028

Table 9.2 Estimated C_{ESD} for ESD1, ESD2 and ESD3 in Al and Cu interconnects for 4kV full ESD protection [4].

Structures		ESD1	ESD2	ESD3
C_{ESD} (pF)	Al	0.97	0.13	0.10
	Cu	0.84	0.12	0.09

of Cu technology significantly reduces such capacitance by an average 30%. Thirdly, the overall C_{ESD} was greatly reduced when using the compact ESD2 (~85%) and ESD3 (~89%) compared to the traditional ESD1 structure. The benefits of the reduction in C_{ESD} to general RF and MS circuit performance are discussed next. The ESD-induced noise effect was studied for the ESD protection structure self-generated noises in this study. The noise coupling effect due to C_{ESD} is relatively easier to understand, hence, not included in this study. The noise models for basic NMOS, BJT and SCR devices described in Chapter 3 are used for noise analysis in this study.

Three typical circuits were selected in this study to address different aspects of the ESD-to-circuit influences accordingly. To evaluate the impact of ESD-induced parasitic capacitance, C_{ESD}, on global clock signal integrity, a fifteen-stage ring-oscillator circuit, running at 4.7GHz, was designed that has different ESD protection structures, i.e., ESD1, ESD2 and ESD3, connected to the I/O pad of the circuit. These ESD protection structures have the same metal coverage and the difference in the C_{ESD} came directly from their different silicon structures in this case. In a single-load case, where an ESD protection structure is connected to only one I/O node of the ring-oscillator circuit, measurements found 85% clock speed reduction due to the C_{ESD} when using ESD1, which can be recovered by around 41% and 62% if using the ESD2 and ESD3 structures, respectively. The on-chip clock corruption problem becomes much more severe in a full-load scenario with ESD protection units connected to every I/O nodes of the circuit. This example clearly demonstrates the C_{ESD}–caused general RC delay effect on global clock signal integrity.

To illustrate the whole-spectrum of the influences of parasitic C_{ESD} on circuits, a low-power, high-performance Op Amp circuit was designed and used as a test vehicle in a second study. The Op Amp circuit features low power (0.43mW), high slew rate (116mV/ns), short settling time (3.7ns at 1%), wide output swing (0,96V at 80% gain) and large unity-gain bandwidth

(121MHz). Figure 9.9 shows the Op Amp schematic consisting of a differential pair input stage for better noise rejection; a source-follower gain stage for high gain and level shift; a class AB complementary push-pull output stage with low quiescent current for high-swing, low power consumption, as well as crossover distortion elimination; and a compensation capacitor C_C with active nulling resistor for wide bandwidth and adequate stability. The three different ESD protection structures, ESD1, ESD2 and ESD3 are connected to the output pad to provide complete ESD protection. The parasitic C_{ESD}, including the parasitic capacitances from both silicon structures and different ESD metal coverage, i.e., using both Al and Cu interconnects, was included in circuit analysis and simulation. Typical circuit characteristics for the Op Amp circuit with different C_{ESD} loads, i.e., C_{ESD}=0 for no ESD protection case and different C_{ESD} values for using ESD1, ESD2 and ESD3, are evaluated, which include the phase Bode plot shown in Figure 9.10, the large signal step response for slew-rate test given in Figure 9.11, and the small signal step response for settling time extraction depicted in Figure 9.12. Comparison data for the circuit specifications are summarized in Table 9.3, from which the significance of the ESD-to-circuit

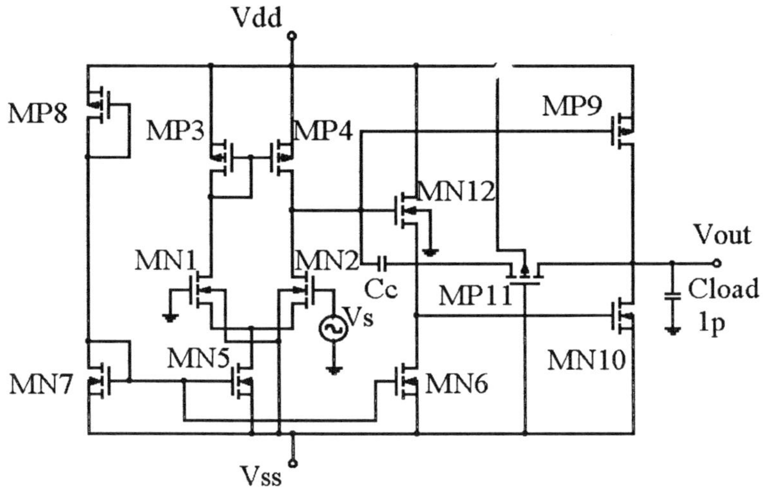

Figure 9.9 A schematic for the high-performance Op Amp circuit implemented in a 0.18μm CMOS process technology [4].

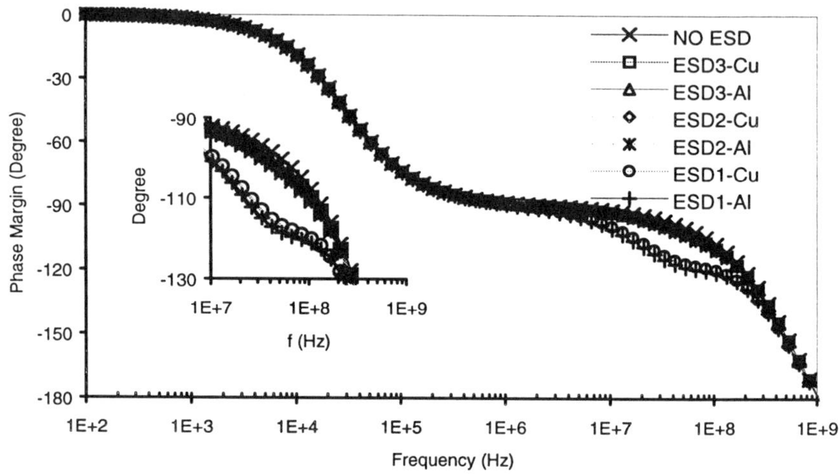

Figure 9.10 Phase Bode plots for the Op Amp circuit with different ESD protection units in both Al and Cu interconnects show the impact of C_{ESD} on the phase margin specification [4].

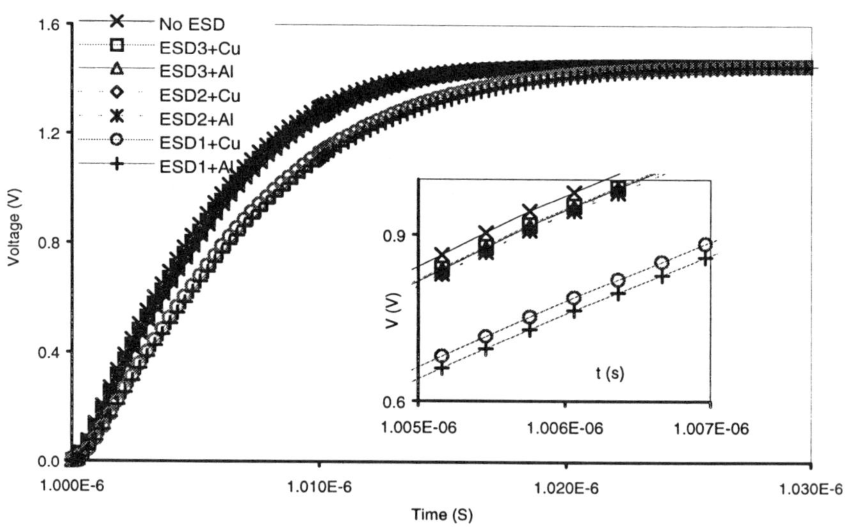

Figure 9.11 Large signal slew rate characteristics for the Op Amp circuit using different ESD protection units in Al and Cu interconnects show the C_{ESD} impact on the response time [4].

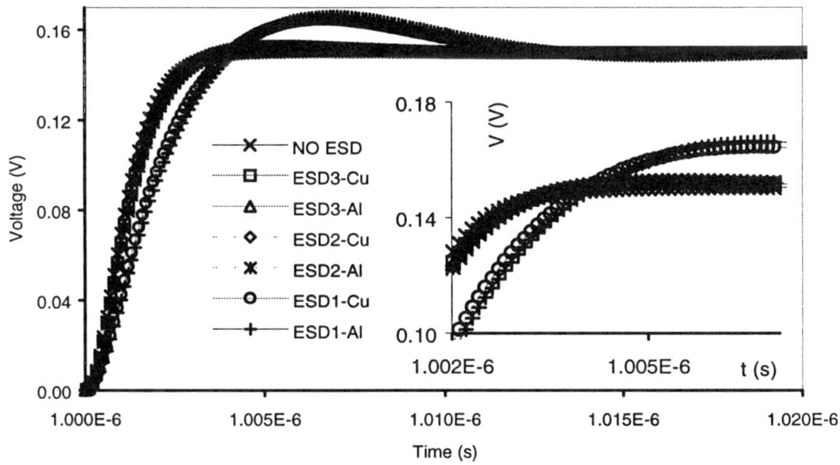

Figure 9.12 Small signal settling time characteristics for the Op Amp circuit using differing ESD protection units in Al and Cu interconnects show the C_{ESD} impact on the settling time response [4].

Table 9.3 A summary for the C_{ESD}-to-circuit influences on the Op Amp circuit using different ESD protection structures in Al and Cu interconnects (data extracted at a load of C_L = 1pF) [4].

		f_T (MHz)	Phase Margin	Slew Rate (mV/ns)	Settling Time (ns)
No ESD		120.7	70.1°	115.9	3.77
ESD1	Al	74.0	60.0°	81.0	17.07
	Cu	77.5	61.2°	84.4	11.92
ESD2	Al	109.9	68.7°	109.9	7.60
	Cu	110.7	68.8°	110.4	7.47
ESD3	Al	112.2	69.0°	111.1	7.15
	Cu	113.0	69.1°	111.5	6.85

influence is clearly observed. For example, the unity-gain bandwidth, f_T, reduces 38.7% when using ESD1 in Al. However, use of compact ESD2 and ESD3 recovers the lose substantially, by 81%. The phase margin decreases by 14% when using ESD1 in Al, which was recovered by 89% when using ESD3. Almost all critical circuit specifications suffer degradation when using the large-size ESD1 structure, which were significantly recovered by using the compact ESD2 and ESD3 protection devices. Further, more improvement was observed in circuits using new Cu interconnect due to even less ESD-metal-induced capacitances. A detailed comparison for the degree of the ESD-to-circuit influences using different ESD protection structures is summarized in Table 9.4. The results clearly show that ESD-induced parasitic capacitances are becoming intolerable to high-performance RF and MS ICs, and compact, robust ESD protection solution are highly desirable in avoiding such circuit performance degradation while maintaining adequate ESD protection.

Table 9.4 A data comparison for C_{ESD}–caused performance degradation of the Op Amp circuit in Al interconnect [4].

Parameters	No C_{ESD}	ESD1	ESD2	ESD3
f_T (MHz)	120.7	-38.7%	-8.9%	-7.0%
			Recovery	
		→	+ 81.9% (+83.5% in Cu)	→
Phase Margin	70.1°	-14.4%	-2.0%	-1.6%
			Recovery	
		→	+ 88.9% (+90.3% in Cu)	→
Slew rate (mV/ns)	115.9	-30.1%	-5.2%	-4.1&
			Recovery	
		→	+ 86.4% (+88.7% in Cu)	→
t_{set} (ns, 1%)	3.77	-353%	-102%	-89.7
			Recovery	
		→	+74.6% (+76.9% in Cu)	→

In a third case study, a low-power 2.4GHz low noise amplifier (LNA) circuit for wireless communication application was designed and implemented in a 0.18μm CMOS process technology to investigate impacts of ESD-self-generated noises on circuit noise performance. This fully integrated LNA circuit is shown in Figure 9.13, featuring a two-stage topology to realize high gain. Current sharing scheme was used in biasing to ensure low power consumption. On-chip inductors and inductive source degeneration technique were used for on-chip impedance matching (50Ω) at

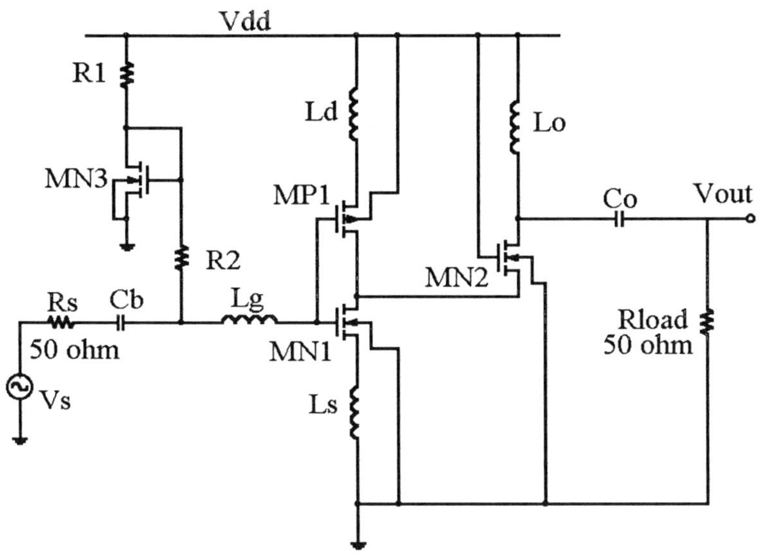

Figure 9.13 Schematic of a low-power 2.4GHz LNA circuit designed in a 0.18μm CMOS technology [4].

both input and output ports. Typical circuit specifications include centre frequency of 2.4GHz, noise figure (NF) of 1.76dB and very low power consumption of 24mW. ESD protection for I/O pad is provided by three different ESD protection structures, i.e., ESD1, ESD2 and ESD3. Generally, there are two types of noise effects associated with ESD protection structures: the first one is the commonly understood substrate noise coupling via C_{ESD}, which can be readily evaluated by injecting noise signals into GND pads, however, was not included in this study; the second one, more important but largely overlooked, is the ESD-self-generated noises, which is directly related to the features and sizes of the ESD protection structures used. This ESD self generated noise effect was the focus of this study. The ESD noise models for the three ESD protection structures, as described before, are included in circuit analysis to evaluate ESD-induced noise impacts on circuit noise performance. The results of the typical S-parameter characteristics for the LNA circuit are shown in Figures 9.14 and 9.15, with critical specifications summarized in Table 9.5. The LNA features include an S_{21} of 23.4dB at the centre frequency of 2.41 GHz and a noise figure NF of

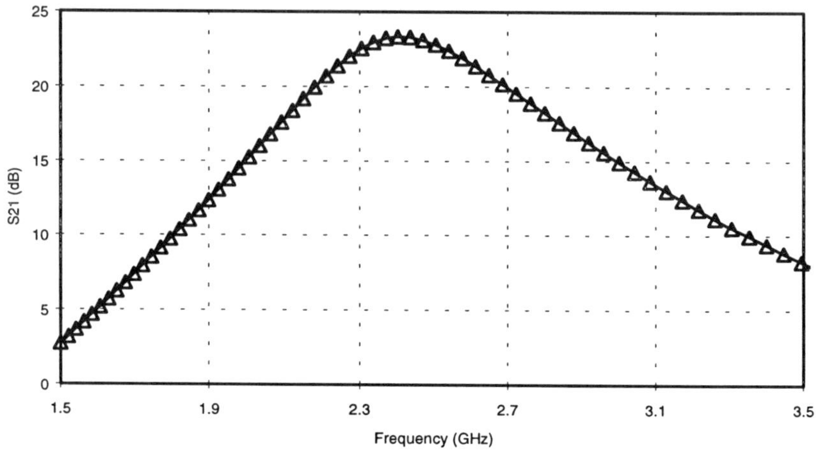

Figure 9.14 S_{21} characteristic of the 2.4 GHz LNA circuit [4].

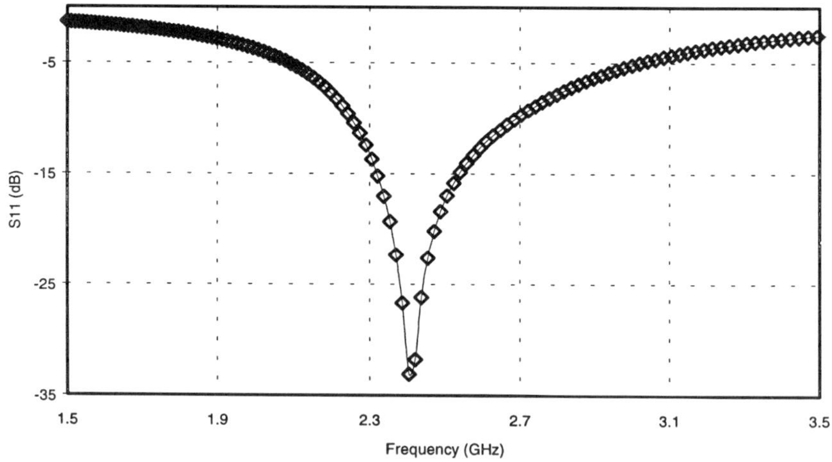

Figure 9.15 S_{11} characteristic of the 2.4GHz LNA circuit [4].

Table 9.5 LNA circuit specifications [4]

Centre Frequency (GHz)	2.41 GHz
S21 (dB)	23.4
S11 (dB)	-34.5
S22 (dB)	-47.7
S12 (dB)	-39.6
Power supply (V)	3
Current (mA)	8.52
Noise figure (dB)	1.76

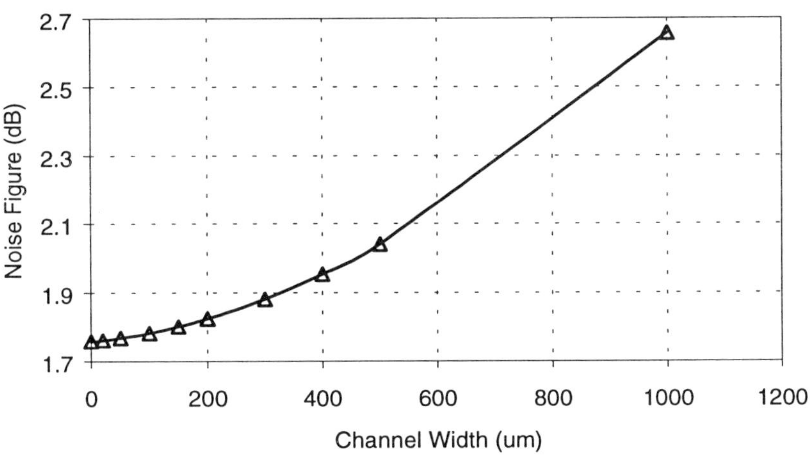

Figure 9.16 Noise figure of the LNA circuit using ESD1 protection with different sizes shows dependence of LNA NF on the ESD-induced noises [4].

1.76 dB. The total current is 8.52 mA at 3V operation, reflecting a very low power consumption at the given supply voltage. Comprehensive evaluation of ESD-self-generated noise contribution to LNA circuit was conducted for the LNA circuit using the ESD protection units, ESD1, ESD2 and ESD3, respectively. Figure 9.16 shows the relationship between LNA NF and ESD protection device sizes using ESD1, i.e., the traditional ggNMOS structure, which clearly indicates the strong ESD-induced noise impact on overall LNA circuit noise performance. Comparison data summarized in Table 9.6 indicates that a traditional ggNMOS (ESD1) structure has noticeably worsened the LNA noise figure (by ~3.78%) due to its large ESD noise generation However, using compact ESD protection, (ESD2 and ESD3) almost totally recovered the degradation in noise performance (a mere ~0.02% increase in NF using ESD3). This implies that ESD-induced noises are not trivial when using conventional ESD protection structures and novel ESD protection solutions are essential to avoiding such circuit noise performance degradation while maintaining adequate ESD protection.

Table 9.6 Impact of ESD self generated noise on LNA noise performance [4]

ESD Protection Type	NF (dB)	Degradation
No ESD	1.7582	-
ESD1	1.8247	3.78%
ESD2	1.7596	0.08%
ESD3	1.7586	0.02%

The above examples demonstrate that the ESD-induced parasitic effects are inevitable and have strong negative impacts on circuit performance. To alleviate the ESD-to-circuit influences, low-parasitic compact ESD protection solutions are highly desirable both to eliminate the ESD-to-circuit influences and to reduce ESD-related Si consumption, particularly, for RF and MS ICs in VDSM regime.

9.4. SUMMARY

In summary, it is important for IC designers to understand that ESD protection circuit design is not about designing individual ESD protection structure itself. A working stand-alone ESD protection structure does not guarantee a functional IC chip. There exist strong ESD-circuit interactions. On one hand, the core circuit may affect the performance of ESD protection significantly, resulting in pre-mature ESD failures due to parasitic internal discharging structures. On the other hand, ESD protection structures have inevitable parasitic effects that influence the functionality of the core circuit substantially, including global clock signal integrity, almost all key circuit specifications and noise performance. This theory has been proved by practical circuit design examples. The solution to the ESD-circuit interaction problem is to design novel, low-parasitic, compact ESD protection structures. It is therefore imperative for IC designers to adopt a system approach in developing novel ESD protection solutions for any IC chips.

REFERENCES

[1] Feng, H. G., "A Mixed-Mode Simulation-Design Methodology for on-Chip ESD Protection Design", *MS Thesis*, May, 2001.

[2] Feng, H. G., Gong, K. and Wang, A., "ESD Protection Design Using Copper Interconnects: More Robustness and Less Parasitics", *SRC Publication: 2000 Publications in Copper Design Challenge, Pub. P000375,* 17-Mar-2000.

[3] Gong, K., Feng, H. G., Zhan, R. Y. and Wang, A. Z., "ESD-Induced Circuit Performance Degradation in RFICs", in press, *Microelectronics Reliability, Elsevier Science,* Oct. 2001.

[4] Gong, K., Feng, H. G., Zhan, R. Y. and Wang, A. Z., "A Study of Parasitic Effects of ESD Protection on RF ICs", in press, *IEEE Trans. Microwave Theory and Techniques,* Feb. 2002.

[5] Gong, K., "ESD Protection in Copper Interconnect and ESD-to-Circuit Performance Influences", *MS Thesis,* May 2001.

Chapter 10

CONCLUSION REMARKS AND FUTURE WORK

10.1. CONCLUSION REMARKS

On-chip ESD protection circuit design is an integral part of development of almost every single IC chip and becomes a main design challenge to the IC design community as IC technology advances into the VDSM regime and beyond. For successes of on-chip ESD protection circuit development, it is imperative for IC circuit designers to thoroughly understand the basics of on-chip ESD protection theory, including ESD failure mechanisms, ESD modelling circuits, ESD standards and regulations, device physics of basic ESD protection elements, and operation of existing ESD protection sub-circuits. It is equally important for IC designers to take a system approach in developing new ESD protection solutions, to follow a simulation-based ESD design methodology for design prediction, and to take into consideration the complex ESD-circuit interactions in practical design.

ESD phenomenon originates from transfer of static electric charges between two objects with different electrostatic potentials as they are brought into contact. The resulting large voltage and current surges may cause severe damages to electronic devices involved. The two types of ESD-induced damages are the catastrophic failures, which are either thermal damages due to material melting or dielectric rupture due to excess electric field, and the latent failures, which cause lifetime reliability problem. Hence, ESD protection measures are required to protect IC devices from being damaged by ESD transients, of which on-chip ESD protection structures are the most efficient means. Depending upon the origins of ESD pulses, different ESD model circuit and standards are developed, including HBM model, MM model, CDM model, IEC model and TLP model, etc, to name a few. The ESD robustness of an IC part is evaluated by conducting ESD stressing tests using an ESD zapping tester and rated in terms of the voltage level it passes in ESD zapping tests, e.g., 2kV, etc. The basic working principle of an ESD protection structure is two-fold: to provide a low-impedance conducting path to discharge the ESD current transient safely to prevent thermal damages and to clamp the pad voltage to a sufficiently low level to avoid dielectric rupture. Several basic semiconductor devices have the right features for ESD current shunting and voltage clamping, including diodes, bipolar junction transistors, MOSFET transistors, and SCR devices. Since different IC devices have different applications and requirements, relatively complicated ESD protection circuits using the basic ESD discharging devices are developed to meet the specific needs, which is particularly the case for RF and mixed-signal ICs in VDSM regime. While traditional trial-and-error approach has been dominating in ESD protection design practices, it is highly desirable to develop simulation-based predictive ESD protection circuit design methodologies in order to achieve the first-time design success and to reduce the development costs. There are many existing ESD simulation approaches currently, including TCAD device level simulation and ECAD circuit level simulation. Practical design examples demonstrate that a mixed-mode ESD simulation-design methodology, involving the complex process-device-circuit-electro-thermal coupling effects, serves as a reasonably good predictive ESD protection circuit design methodology, should proper simulation calibration be performed during the simulation-design course. More reliable and efficient ESD simulation methods are currently under research. It is very critical for IC designers to understand that there exist complex interactions between the ESD protection units and the core circuit being protected. On one hand, the core circuits may affect the ESD performance at chip level, often leading to pre-mature ESD failures due to weak parasitic discharging links inside the core circuit. One the other hand, ESD protection structures produce parasitic effects inevitably, including ESD-induced parasitic capacitances and ESD-related noises. The parasitic ESD effects may have substantial influences on performance of the core circuits being protected, particularly for RF, mixed-signal and high pin count high-density VLSI chips. Therefore, it is imperative for IC designers to treat an ESD protection design task as a full chip design work and to verify the final design at whole chip level.

It is worth to repeat that ESD protection structure is not portable either across different IC products in the same process family or between IC chips developed in different process technologies. It will be unacceptable to continue rely on experience-based ESD protection design approaches in designing ESD protection circuits in near future. Mixed-signal ESD simulation and whole-chip ESD design verification must be used in future advanced ESD protection circuit development in order to ensure design successes, to reduce development costs and to shorten the critical time to market. It is also important to point it out that, while this book introduces many ESD design methodologies and provides a number of practical ESD protection circuit design examples, one should not try to duplicate those things in your own design practices. One could never over-emphasis the importance of creative thinking and vigorous quantitative verification in practical ESD protection circuit design.

10.2. FUTURE WORK

While significant progresses have been made in the field of on-chip ESD protection circuit design in recent years, far more efforts in research are needed to make ESD protection circuit design task a predictive, reliable and efficient one. It would certainly be unwise to try to predict the future of ESD protection circuit design, nevertheless, the following problems emerge as major challenges in ESD protection circuit design. Firstly, a whole chip ESD protection design verification methodology and software package are in urgent need. The end users should be those ordinary IC circuit designers, instead of the ESD protection design specialists. The design methodology should provide a rational way for IC designers to conduct the complete ESD protection simulation all the way to the chip level before running any silicon. While design experience is valuable, it is the quantitative simulation that can guarantee the final design success. CAD tools providing whole chip design verification functions, which can address the complex ESD-circuit interactions at chip level, are essential in future ESD protection circuit design. Such CAD-based ESD design verification can be divided into two parts, i.e., layout level and schematic level. The layout level ESD design verification should include not only the normal DRC and LVS checking, but also parametric checking, which is a smart physical checking function that is able to extract all the ESD-oriented parasitic devices, to extract the critical ESD parameters, such as trigger threshold, holding point, discharging impedance and gain, etc, to reduce the head counts of the extracted ESD devices, and to examine the extracted ESD device netlist against the design requirement according to the critical ESD parameters. At schematic level, comprehensive electro-thermal simulation should be conducted to verify both ESD protection operation and the core circuit functionality in order to guarantee a functional chip. It is critical to take into consideration of the ESD-circuit interactions at the whole chip level. Secondly, in order to realize full-chip ESD design verification, accurate ESD device models must be developed, which address key features such as high current operation, avalanche snapback behaviours, ESD failure criteria for both thermal and dielectric damages, geometry dependence of ESD damages, and latent ESD damages, etc. A parallel, section-wise ESD device modelling technique may serve to address the geometry-dependent ESD failure problem. Thirdly, innovative compact and low-parasitic ESD protection solutions must be developed for future advanced IC chips, particularly for RF, mixed-signal and high-density ICs in VDSM regimes and beyond. Finally, ESD protection for nano-scale technologies must be investigated for future IC technologies. None of the above tasks is an easy one. However, this is a challenge an IC circuit designer must face, like it or not. For now, the one positive thing is that, so far, nothing has stopped the IC technologies from be advancing continuously yet.

Appendix A

SUMMARY FOR ESD TEST STANDARDS

Table A1 A summary for different HBM ESD test standards

Standards	Year	Application	Model elements	Set-up	Waveform verification				Zapping procedures		
					Load	Voltage	Rise time t_r(ns) (short)	Decay t_d(ns) (short)	Zap times	Zap intervals Δt	Sample size
MIL-STD-883E Method 3015.7 [1]	1989	Device	$R_{ESD}=1500\Omega$ $C_{ESD}=100pF$	Socketed	Short	±4000V	<10	150±20	3x +/- each pin	1s	3
ESD STM5.1-1998 [2]	1998	Device	$R_{ESD}=1500\Omega$ $C_{ESD}=100pF$	Socketed	Short/ 500Ω	±250V-8000V	2-10	150±20	1x +/-	0.3s	>3
JESD22-A114-B [3]	2000	μ-circuit	$R_{ESD}=1500\Omega$ $C_{ESD}=100pF$	Socketed	Short/ 500Ω	±250V-4000V	2-10	130-170	1x +/-	> 0.3ms	3
AEC-Q100-002 REV-C [4]	1998	IC	$R_{ESD}=1500\Omega$ $C_{ESD}=100pF$	Socketed	Short/ 500Ω	±1000V /4000V	2-10	130-170	1 or 3 x +/-	> 0.5s	3
IEC 1000-4-2 [5]	1995	Equipment	$R_{ESD}=330\Omega$ $C_{ESD}=150pF$	-	Short	±1000V /8000V	0.7-1ns	60 @ 26.6%I_p	>10x	1s	-

Table A2 A summary for different MM ESD test standards

Standards	Year	Application	Model elements	Set-up	Waveform verification				Zapping procedures		
					Load	Voltage	Rise time t_r(ns) (short)	Ring f/t (short)	Zap times	Zap intervals Δt	Sample size
ESD STM5.2-1999 [6]	1998	Device	$R_{ESD}=0\Omega$ $C_{ESD}=200pF$	Socketed	Short/ 500Ω	±100V – 400V	~7.5	80ns	3x +/-	1s	>3
JESD22-A115-A [7]	1997	μ-circuit	$R_{ESD}=0\Omega$ $C_{ESD}=200pF$	Socketed	Short/ 500Ω	±400V	~6	11-16MHz	1x +/-	>0.5s	3
AEC-Q100-003 REV-E [8]	2001	IC	$R_{ESD}=0\Omega$ $C_{ESD}=200pF$	Socketed	Short/ 500Ω	±400V	~6	66-90ns	3 x +/-	> 1s	3

Table A3 A summary for different CDM ESD test standards

Standards	Year	Application	Model elements	Set-up	Waveform verification				Zapping procedures		
					Load	Voltage	Rise time t_r(ps) (Disc)	HH-width (Disc)	Zap times	Zap intervals Δt	Sample size
ESD STM5.3.1-1999 [9]	1999	Device	Minimizing parasitics	Socketed	Cu Discs: 4/30pF	±500V	<200/4pF <250/30pF	<400ps/4pF <700ps/30pF	3x +/-	1s	>3
JESD22-C101-A [10]	2000	μ-circuit	Minimizing parasitics	Non-socketed	Cu Discs: 6.8pF	±500V	400	1ns ±0.5	5x +/-	>0.2s	3
AEC-Q100-001 REV-A [11]	2001	IC	Minimizing parasitics	Socketed	Cu Discs: 4/30pF	±250V - 2000	<400	<600ps	3 x +/-	> 1s	3

REFERENCES

[1] *MIL-STD-883E, Method 3015.7*, Electrostatic Discharge Sensitivity Classification, Dept. of Defence, Test Method Standard, Microcircuits, 1989.

[2] *ESD STM5.1-1998 Revised*, Electrostatic Discharge Sensitivity Testing – Human Body Model Component Level, the ESD Association, February 1998.

[3] *JEDEC JESD22-A114-B*, Electrostatic Discharge Sensitivity Testing Human Body Model, the Electronics Industries Alliance, June 2000.

[4] *AEC – Q100-002 Rev-C*, Human Body Model Electrostatic Discharge Test, the Automotive Electronics Council, October 1998.

[5] *IEC 1000-4-2*, "Electromagnetic Compatibility, Part 4: Testing and Measurement Techniques, Section 2: Electrostatic Discharge Immunity Test", The International Electrotechnical Commission, January 1995.

[6] *ESD STM5.2-1999 Revised*, Electrostatic Discharge Sensitivity Testing – Machine Model Component Level, the ESD Association, May 1999.

[7] *JEDEC JESD22-A115-A*, Electrostatic Discharge Sensitivity Testing Machine Model, the Electronics Industries Alliance, October 1997.

[8] *AEC – Q100-003 Rev-E*, Machine Model Electrostatic Discharge Test, the Automotive Electronics Council, January 2001.

[9] *ESD STM5.3.1-1999 Revised*, Electrostatic Discharge Sensitivity Testing – Charged Device Model Component Level, the ESD Association, September 1999.

[10] *JEDEC JESD22-C101-A*, Field-Induced Charged-Device Model Test Method for Electrostatic Discharge-Withstand Thresholds of Microelectronic Components, the Electronics Industries Alliance, June 2000.

[11] *AEC – Q100-011 Rev-A*, Charged Device Model Electrostatic Discharge Test, the Automotive Electronics Council, January 2001.

Appendix B

COMMERCIAL ESD TESTING SYSTEMS

Table B1 Popular ESD testing system models from Thermo KeyTek (http://www.keytek.com, data extracted in October 2001)

Vendor Models	ESD Test Standards
RCDM	**CDM Model:**
	JEDEC C101
	ESD STM5.3.1
PARAGON	**HBM Model:**
	MIL-STD-883D, 3015.7
	ESD STM5.1 -1998
	JEDEC JESD22-A114
	MM Model:
	ESD STM5.2
	JEDEC JESD22-A115

Vendor Models	ESD Test Standards
WaferMaster	**HBM Model:**
	MIL-STD-883D, 3015.7
	MM Model:
	MM, 200pF, 0Ω
ZAPMaster 7/6	**HBM Model:**
	MIL-STD-883D, 3015.7
	ESD STM5.1
	MM Model:
	EIA/J
	CDM Model:
	Socketed CDM model

Table B2 Popular ESD testing system models from Oryx Instruments Corp. (http://www.orysinstruments.com, data extracted in October 2001)

Vendor Models	ESD Test Standards
System 700	**HBM Model:** MIL-STD-883D, 3015.7 ESD STM5.1 JEDEC JSED22 A114
	MM Model: JEDEC A115
	IEC Model: IEC-801-2
Aurora	**HBM Model:** MIL-STD-883D, 3015.7 ESD STM5.1 JEDEC JESD22-A114
	MM Model: ESD STM5.2 JEDEC JESD22-A115 EIA/J
Orion	**CDM Model:** ESD STM5.3.1 JEDEC JESD22-C101
11000EX	**HBM Model:** MIL-STD-883D, 3015.7 ESD STM5.1 JEDEC JESD22-A114
	MM Model: ESD STM5.2 JEDEC JESD22-A115 EIA/J
	TLP: (Optional)

Appendix C

ESD PROTECTION CIRCUIT DESIGN CHECKLIST

While quantitative analysis and simulation are essential to successful ESD protection circuit design, some simple design checking work is very beneficial in practical design. The following checklist serves to assist IC designers in designing ESD protection circuits.

- Understand that ESD protection solution is not portable, neither among different IC products in the same process technology nor between different process technologies.

- When borrowing a successful ESD protection structure, make sure to understand its operation. A specific IC circuit may affect the ESD protection structure significantly.

- Check to make sure every pad has at least one ESD protection structure connected, depending upon the whole-chip ESD protection scheme adopted, excluding the self-protection output pads.

- Check to make sure a low-impedance discharging path existing between any two pads.

- Check to make sure a designed discharging path functioning for ESD transients of different polarities, according to the ESD test standard applied.

- For any ESD protection structure, accurately define the I-V characteristic parameters, such as triggering point (V_{t1}, I_{t1}), holding point (V_h, I_h), and thermal failure threshold point (V_{t2}, I_{t2}).

- Estimate the discharging channel resistance to make sure the size of the ESD protection structure is adequate for the targeted ESD protection level.

- Check to make sure the V_{t1} of a power clamp is set with adequate safe margin (~50% higher than V_{DD}) to avoid accidental turn-on by normal power supply fluctuations.

- Check to make sure the V_h is low enough to avoid gate oxide overstress.

- Check to make sure the $V_{t1}<V_{t2}$ condition is satisfied if a multi-finger NMOS ESD protection structure is used to ensure uniform turn on.

- Check to make sure the ESD protection structure responds to ESD pulses quickly enough, i.e., $t_1 \leq t_r$ for the HBM, CDM or IEC model to be used.

- If a primary-secondary protection scheme is used, check to make sure the isolation resistor is large enough to limit ESD current flowing into the core circuit and include the R in circuit simulation.

- Check all resistors used in ESD protection network for their values and layout. Use smoothed deep-diffusion heads for resistors.

- If poly-Si resistors are used, check the current handling capacity.

- For multi-finger ESD structure, properly arrange the current flow paths in layout to avoid current crowding.

- For NMOS ESD structures, check the DCGS and SCGS for proper values. Pay special attention to the minimum SCGS traditionally suggested if sub-0.25μm technologies are used.

- For gcNMOS ESD structure, check the gate bias V_G to make sure no overstress occurring at the gate.

- For SCR type ESD structures, check the latch-up immunity and make sure the I_h is higher than any other on-chip current (i.e., I_{DD}) to avoid staying in latch-up state after ESD transients.

- Check the spacing between ESD devices and any internal devices that produce substantial current (i.e., charge pumps & output buffers), since such large current may leak into the ESD region via substrate to cause early triggering. Use guard rings to de-couple the substrate current leakage.

- Check all diffusion layer spacing in layout to avoid any parasitic transistors that often lead to pre-mature ESD failures.

- Check to make sure the metal interconnects are wide enough to survive ESD pulses. However, be cautious in using over-size metal line because of its parasitic RC effects. Conducting simulation if possible.

- Check ESD metal routing to make sure the shortest and easiest current flowing paths in order to avoid current crowding and overheating effects.

- Check the current handling capacity of contacts and vias. Use as many of them as possible.

- Check the placement of contacts and vias to make sure uniform current pick-up across the ESD protection structure.

- Avoid using interconnect bridges within ESD protection network if ever possible.

- Check all possible parasitic ESD-type devices as potential weak discharging links to suppress any possible circuit-to-ESD influences that often causes pre-mature ESD failures.

- To evaluate the ESD-to-circuit influences including RC delay, noise coupling through ESD protection structures and self-generated noises by the ESD protection structures, and include these parasitics into circuit simulation.

- Make sure to use the salicide-blocking mask in the technologies with salicide feature.

- Make sure to use the LDD-blocking mask in LDD CMOS technologies.

- To use compact ESD protection structures if ever available, particularly in high-speed dense design.

- To evaluate the leakage current level of ESD protection structures.

- In mixed-signal IC design, try to estimate the dV/dt or dI/dt associated with the large digital noises to make sure no accidental triggering of the ESD protection structures.

- In RF IC design, try to estimate the dV/dt or dI/dt associated the high-frequency signals to make sure no accidental triggering of the ESD protection structures.

Index